THE FLORIDA STRAWBERRY FESTIVAL

A Brief History

Fresh Squeezed
Ice Cold
CITY

THE FLORIDA STRAWBERRY FESTIVAL

A Brief History

By Lauren McNair and Gil Gott

The Donning Company Publishers
731 South Brunswick Street
Brookfield, MO 64628

Lex Cavanah, General Manager
Nathan Stufflebean, Production Supervisor
Anne Burns, Editor
Rick Boley, Graphic Designer
Jennifer Elam, Project Research Coordinator
Katie Gardner, Marketing and Project Coordinator

Lynn Walton, Project Director

THE
DONNING COMPANY
PUBLISHERS

Library of Congress Cataloging-in-Publication Data

Names: McNair, Lauren, author. | Gott, Gilbert, author.
Title: The Florida Strawberry Festival : a brief history / by Lauren McNair and Gilbert Gott.
Description: Brookfield, MO : The Donning Company Publishers, [2017] | Includes bibliographical references.
Identifiers: LCCN 2016057795 | ISBN 9781681841038
Subjects: LCSH: Florida Strawberry Festival--History. | Agricultural exhibitions--Florida--Plant City--History. | Food festivals--Florida--Plant City--History. | Fairs--Florida--Plant City--History. | Strawberries--Florida--Plant City.
Classification: LCC S555.F82 F586 2017 | DDC 630.74/75965--dc23
LC record available at https://lccn.loc.gov/2016057795

Printed in the United States of America at Walsworth

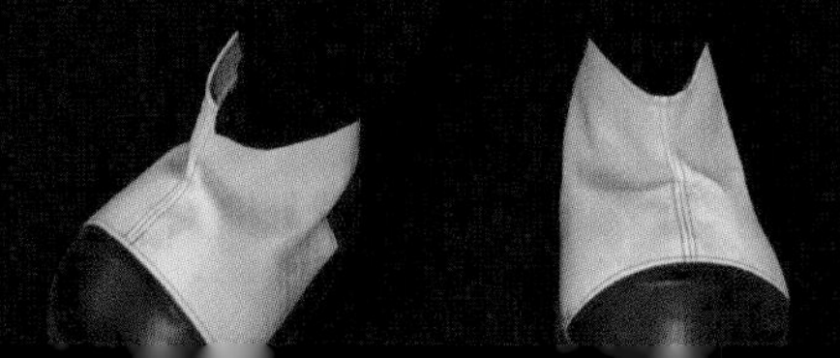

CONTENTS

PREFACE:
How This Story is Put Together and What You Will Find Here 7

ACKNOWLEDGMENTS:
A Word of Gratitude 11

INTRODUCTION:
Plant City and the Festival—How They Came to Be 13

Chapter One:
THE FLORIDA STRAWBERRY FESTIVAL—AN OVERVIEW 18

Chapter Two:
THE FLORIDA STRAWBERRY FESTIVAL—THE EARLY YEARS
by Gil Gott 22

The 1930s: The Formative First Years 24
The 1940s: A New Decade Begins, Dims, and is Rekindled 45
The 1950s: A Decade of Postwar Normalcy and Growth 54
The 1960s: A Changing of the Guard, New Leaders, and New Ideas 70

Chapter Three:
THE FLORIDA STRAWBERRY FESTIVAL—MODERN TIMES
by Lauren McNair 92

The 1970s: The Fifth Decade of Festivals 94
The 1980s: The Sixth Decade of Festivals 111
The 1990s: The Seventh Decade of Festivals 127
The 2000s: The Eighth Decade of Festivals 141
The 2010s: The Ninth Decade of Festivals 154

APPENDICES:
The Organizers, Queens, and Many Others Who Made It Happen
I. Charter Members 170
II. Queens and Courts 171
III. Presidents 185
IV. Directors 188

BIBLIOGRAPHY 193

INDEX 194

ABOUT THE AUTHORS 208

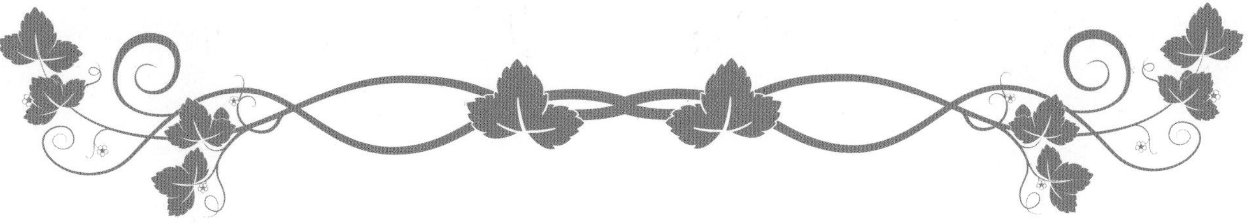

PREFACE:
HOW THIS STORY IS PUT TOGETHER AND WHAT YOU WILL FIND HERE

Before welcoming you to our brief history of the Florida Strawberry Festival we want to tell you what you are about to see. The planning for the strawberry festival began in 1929, during what is known as the Great Depression. Yet, Plant City area farmers, merchants, businesspeople, and townspeople were excited about the opportunity to show their excellent produce and their beautiful town. They devoted themselves to producing a highly successful festival. And they did. On Wednesday, March 12, 1930, the first strawberry festival opened to a crowd of 15,000!

It continued on to today, with a six-year break during World War II, growing year after year. There is really too much to tell you than what we could comfortably fit into a book this size. Since we cannot cover all the Grand Parade floats, bands, marching units, and all the many vendors, the Baby Parade, the agriculture exhibits, plant exhibits, art, and folklore, we packaged the story in portions.

To begin, we will first give you a good overview of Plant City. When we say Plant City we mean the town and the surrounding communities such as Springhead, Hopewell, Cork, Dover, Seffner, Trapnell, Lithia, Turkey Creek, and all the others that make up this outstanding community. You see, Plant City is very different from many other cities and we will tell you about that in our introduction.

In Chapter One: The Florida Strawberry Festival—An Overview, we will provide a very brief look at the festival itself—what it consists of and generally how it developed over the years to become what it is today. Although many specific details may be omitted from this section, the reader may be able to acquire information about certain events and activities from elsewhere in this book or by contacting the Florida Strawberry Festival Association or the Plant City Photo Archives and History Center.

Chapter Two: The Florida Strawberry Festival—The Early Years and Chapter Three: The Florida Strawberry Festival—Modern Times will take you through the years individually from the festival's founding in 1930 to the recent years, with an initial emphasis on 1930 and the wonderful inauguration of this fascinating festival. Here, you will find the many beaming queens, the Grand Parade and its many floats, some details of the Baby Parade, the midway, the entertainment, the agricultural exhibits, art and garden shows, home recipes, canning, baking, and the many and varied vendors.

We want to mention that written accounts of what took place on the days of the festival are sketchy, if available at all. Since most of the major attractions were held outdoors, we usually don't know what the weather was like and what effect there may have been on the scheduling. Normally in March, the temperature can vary from the thirties to the eighties, and sometimes the cold rains come fast upon a chilly evening. Also, scheduling of dignitaries such as the governor was subject to the duties of the office, and adjustments had to be made. We are using the printed programs to tell what was supposed to take place, and unless there are published accounts of what happened, which we occasionally have access to, we present the planned activities.

Chapters Two and Three cover eighty-six years and eighty spectacular strawberry festivals, and here we'll also talk about the pre-parade luncheons, the fashion show, the Strawberry Ball, and the American Legion raffle. The roles of the Lions Club, the American Legion, the Woman's Club, the Junior Woman's Club, city officials, the thousands of volunteers, and the Florida Strawberry Festival Association officers and directors, without whom none of this would have taken place, are also highlighted.

The appendices include the many queens and court members, the presidents and managers, and the many directors who have given freely of their time over the past decades. And, of course, we love and appreciate the thousands of volunteers who make possible the production of this community-wide annual celebration of the delightful strawberry and all it means to the people of the Plant City area.

We invite you to enjoy the book and please feel free to contact either the Florida Strawberry Festival Association or the Plant City Photo Archives and History Center with your comments and questions.

Florida Strawberry Festival officials welcome Vice President George H. W. Bush to the 1984 festival.

PLANT CITY and
Florida Strawberry
Welcomes
VICE

VALENTINO'S LONDON BROIL
STEAK SANDWICH
LONDON BROIL
Jammer's Farmhouse
Corn Dogs
Cold Drinks Lemonade
French Fries ★ Corn Dogs

ACKNOWLEDGMENTS: A WORD OF GRATITUDE

The authors wish to thank the many people who made this book come to fruition. First we thank the officers and directors of the Florida Strawberry Festival Association for the encouragement and support they have given over the months of putting this together. Second we thank Association General Manager Paul Davis for his passionate support for the preservation of the history of the festival and for bringing together the History Committee to breathe inspiration into this project and also committee members Terry Ballard and Al Berry for the insight and motivation they inspired. We thank the photographers Bill Friend, Gladys Jeffcoat, Billy Friend, and Harry Jeffcoat for many of the wonderful photos you see herein. We thank the staff members at our workplaces for their assistance and for their continuing support. Even when we could not be of much assistance to them, they were there for us.

On a personal note, we gratefully acknowledge the continuing support of our spouses and families who both provided constant encouragement and were very tolerant of the time spent away from the home front.

Last we very enthusiastically thank the staff at the Donning Company Publishers for their continuing work to keep us on the right track and our editor, Anne Burns, for providing great assistance in making a wonderful publication from the piles of drafts we submitted in less than a timely manner.

If there are errors in this book, we accept full responsibility and will welcome any corrections readers may offer. Future revised editions will benefit from your positive critiques.

Thank you all.

INTRODUCTION:
PLANT CITY AND THE FESTIVAL—HOW THEY CAME TO BE

In the early 1800s south Florida barely existed in the minds of most Americans. It was under Spanish rule until 1821 and became a US territory in 1822. At that time Florida ran from St. Augustine to Pensacola, with the center point of Tallahassee becoming the capital. South of that was considered wilderness.

With the Armed Occupation Act of 1842 appealing and offering free land for settlers, Florida became more attractive to agricultural families from Alabama, Georgia, and South Carolina and other adventurous individuals seeking better climate, better and more land, and the promise of opportunities. Three years later, statehood was granted.

The rapid advancement in transportation following the growth of technology, most notably steamboats and railroads, and the result of Florida's Internal Improvement Act in 1855 made migration even more attractive to families looking for a new and better way of life. They came on foot, in wagons, on horseback, via steamboats, and any combination of these to lay claim to fertile farmland and a climate with a greater agricultural productivity.

A major boost in population followed the end of the Civil War. Families from war-ravaged areas in states such as South Carolina, Georgia, Alabama, and Tennessee left their homelands and moved to Florida, now settling in what was called south Florida.

Another area adjacent to what became Plant City was also affected by the close of the Civil War. In late 1865, following emancipation, former slaves from Hopewell, Knights, Springhead, and other areas gathered at the plantation of Sarah Howell in the Springhead area, about seven miles south of Plant City, and established a community later known as Bealsville. They were farmers and became an important segment of the greater Plant City agricultural community over the years.

Railroads provided a new advantage and greater and easier transportation. Villages sprang up along the rail lines throughout the state—and Plant City was one of them. The railroad became one of the greatest advantages the town, and later the city of Plant City, could have dreamed of. Many instant villages owed their existence to the railroad and to its outside capital, which came from the industrialized North.

Henry Bradley Plant, a Connecticut Yankee after whom the town was named, completed the railroad connecting Tampa on the west to Sanford on the east and then to the rest of the world in January 1884. It ran through the area called Hichipucksassa, which then became an instant village. The new village became the town of Plant City, quickly formed and chartered on January 10, 1885, thus beginning the essential relationship between Plant City and the railroad.

Connecticut businessman Henry Bradley Plant (1819–1899) built the railroad that connected Plant City to the world.

In the decade following 1885, Hillsborough County saw a slow evolution from the small agricultural and manufacturing operations to strongly developing industries. This period also saw the growing separation between the manufacturing path Tampa took and the agricultural direction followed by Plant City. By 1886 Plant City had already begun to develop into a center of commerce for east Hillsborough County's agricultural production.

Over the years, the growth of Plant City epitomized the strength of southern rural values. The people were basically rural migrants who carried with them the culture of the countryside—those values found in virtually every aspect of community life. This was Southern Christian

A group of Plant City citizens pose by the busy Union Station, circa 1911.

morality, evangelical Protestantism, individualism, self-sufficiency, chivalry, and a concept of social hierarchy. These and the agricultural work ethic connected these people, developing their own character and a common bond to the place they lived and to the people around them.

With the railroad transportation system, the agricultural sector's productivity dramatically increased, and with that came a steady rise in the commercial sector. Nonetheless, setbacks followed. The Big Freeze of December 1894 and February 1895 almost destroyed all of Florida's citrus production. Some farmers lost their crops, their plants, and their fields, while some merchants, unable to collect on credit extended, went bankrupt. Plant City-area farmers were forced to adapt to changing economic factors. Some turned to strawberries. By 1896 the popularity of strawberries as a major crop soared. By 1902 Plant City stood out as Florida's largest shipping point for strawberries.

Diversification has also been one of the strong points of Plant City's economy. The Warnell Lumber and Veneering Company relocated to Plant City in 1898, employing hundreds of local residents. The Coronet Phosphate Company, established in 1908, followed and it, too, brought many jobs and opportunities to the area. Long a center for logging, lumber, and its products, along with naval stores (turpentine), cattle, and a variety of farm products, Plant City's emphasis turned notably to strawberry production and the industries that support it. In 1920 Plant City claimed the title of "World's Winter Strawberry Capital." With its intersecting rail transportation, it was also the largest inland shipping point in the state of Florida, with fifty to sixty trains on a single day.

In 1921 the newly formed Kiwanis Club urged a revival of the Board of Trade, and its efforts resulted in a new organization. The East Hillsborough County Chamber of Commerce promoted the agricultural and commercial strengths of Plant City. Another boost to the strawberry farmers were the successful efforts by local politicians. In 1925 the state legislature made a special appropriation to the state plant board establishing an agricultural research center. The office and laboratory were set up in the Growers Building on North Palmer Street in downtown Plant City.

In the early 1920s Florida experienced phenomenal growth called the "Land Boom." It didn't last. Florida's real estate boom was called a "classic speculative bubble" and a prelude to the stock market crash. In 1926 Florida's banks crashed. It devastated Florida's economy and drove its recession into a depression. In 1926 alone, bank assets in Florida fell more than $300 million. By 1929 assets fell 60 percent, from $943 million to $375 million. Yet, undeterred, Plant City's optimism continued as it celebrated the opening of the beautiful Hotel Plant in downtown Plant City on November 11, 1926.

Plant City was further boosted in 1927 when Hillsborough County purchased land in Springhead and the state legislature appropriated funds to continue the strawberry program. A laboratory and office were set up in Springhead as the Strawberry Investigation Laboratory. The impact of the depressed economy, however, affected Plant City in many ways. Teachers' pay was cut, class sizes were increased, terms were shortened, and graduation and wedding celebrations were often held at home. The school paper, *The Spokesman*, was discontinued, as was the yearbook, *The Kanyuksaw*. By 1929 two of the three Plant City banks were closed. The city lost merchants and had difficulty collecting taxes.

Plant City's economy, however, was diversified and not dependent on tourism or development alone. Where other cities lost population, Plant City's grew. The abundance of natural resources and the richness of its climate and soil gave it a broad foundation from which to grow. Its agricultural sector continued its productivity, and on July 5, 1929, the newly organized Lions Club came up with the concept of a strawberry festival. By 1930 close to six-thousand acres were in strawberry production and had become Plant City's prominent industry. In one day the railroad shipped thirty cars of berries—518,400 pints of strawberries!

The explosive growth of nearby Tampa, the Great Freeze (1884–1885), the Yellow Fever epidemic (1887–1888), devastating fires (1907 and 1910), World War I (felt in the area from 1917 to 1918), the Florida Land Boom (1920–1925), the Florida Land Bust (1926), the Florida Bank Crash (1926), and the Great Depression (1929–1939) all affected Plant City, yet it survived the trials. The railroads continued their expansion and Plant City experienced forty-four passenger trains daily during the winter tourist season, not to mention the numerous freight trains. The city developed an urban center, providing a structural foundation

Frank DeVane's Springhead strawberry packing shed was a busy place in the 1920s.

The impressive Ferris wheel stands out at the Florida Strawberry Festival fairgrounds.

for economic expansion and stability leading to political and social development resulting in the further sense of community and attachment to place.

With the support of the city and the many outlying communities, the first strawberry festival opened in March of 1930. The strawberry growers brought to market 155,000 quarts of berries on opening day. Fifteen refrigerated boxcars moved out that night. Despite inclement weather, with rain three of the four days of the festival, total attendance for the first festival approximated forty thousand. The "Johnny J. Jones Exposition" continued its midway shows for another week to compensate for the harsh weather. And the festival association crowned its first Strawberry Festival Queen—Charlotte Rosenberg. This was Plant City's response to its many challenges.

The national economy, however, continued in its decline as the Great Depression set in. Fortunately, like communities throughout America, Plant City benefitted from New Deal programs. The projects developed under the Civilian Conservation Corps (CCC), founded in 1933, and the Works Progress Administration (WPA), founded in 1935, kept many residents of the Plant City area working. Included in these projects were drainage for mosquito control, road construction, and the construction of a new National Guard Armory complex (1936–1937). WPA workers prepared Adelson Field in 1935 for the arrival of the Buffalo Bisons baseball team and provided bleachers, a press box, a clubhouse, and floodlights. The WPA also constructed the State Farmers Market in 1939, with an auction shed which was the largest in the state of Florida. Approximately 1,300 to 1,500 farmers used this facility.

The spirit of the berry farmers was boosted by the success of the annual festivals, and Plant City's population continued its growth—now 7,491 by 1940. It was a stable and prosperous community. As the twelfth annual festival opened in 1941, no one could foretell what was soon to come. Jane Langford was crowned queen, and in addition to her local appearances, she rode on the Florida exhibit float in Atlantic City as part of the Miss America Pageant on September 2 and 3. It was heady stuff. Then, on December 7, 1941, the Japanese attacked Pearl Harbor. World War II had begun for the United States.

As Betty Barker Watkins, educator and Plant City's first female commissioner, said that "it was like the sucking of a vacuum that swept away all the men off the streets." It was the war. The Florida Strawberry Festival took a back seat to the demands of the war effort—the service men and women, the home front, guards, rationing, blackout drills. The festival would not come back until the impact of the war was beginning to settle.

In 1947 the American Legion, now with many new and returning members, believed it was time to revive the festival, and once again the community joined in the effort. On March 9, 1948, under the sponsorship of the American Legion and after a six-year hiatus, the Florida Strawberry Festival opened on a ten-acre complex off West Reynolds Street. The festival has opened to happy crowds ever since, growing larger in area and activities and greater in attendance.

The local Norman McLeod Post 26 of the American Legion restarted the Florida Strawberry Festival in 1948 and is shown here at its charity fundraising event in 1949.

- Chapter One -

THE FLORIDA STRAWBERRY FESTIVAL

An Overview

The Florida Strawberry Festival started as a community fair and has remained this way over the many years of its celebration. There have been numerous changes as it has evolved and it remains true to its spirit. It is the community that operates it. Without a volunteer board of directors, literally thousands of volunteers, and great support from the community—all the communities in the greater Plant City and East Hillsborough area—the festival could not exist. The traditions run deep. And the strawberry is still king.

Festival directors recognized the huge diversity in the community and continued adding more exhibits and more free shows and slowly diverted away from the carnival atmosphere and poor reputations many fairs had acquired. Exhibits now include agriculture, commerce, industry, livestock, fine arts, horticulture, and crafts, among others. Again, with the assistance of the Lions Club, the entertainment offerings grew from only local talent to today's top names in country music and other genres including Taylor Swift, Reba McEntire, Little Big Town, and, of course, Plant City native Mel Tillis. Plus, there are free shows throughout the grounds on a daily basis.

The festival grew to an eleven-day exposition and now include the Grand Parade, the Baby Parade, and the Youth Parade. The queen pageant, which has dropped the swimsuit competition, features five finalists of whom one becomes queen, and the contest itself emphasizes the scholarship aspect and additional prizes. The girls have become community ambassadors and appear at many special functions throughout the area and other nearby venues. The Belle City Amusements Midway is regulated and delights many festival goers with rides daily and special events like ride-a-thons. And there is more food and a greater variety than most attendees would expect. As always, there are strawberry shortcakes, strawberry sundaes, strawberry milk shakes, and flats of strawberries to take home.

With annual attendance of 500,000 or more, the Florida Strawberry Festival has developed into not only the best little community fair but one of the best festivals in the country. See you at the festival!

Florida Strawberry Festival Queen Maria Junquera and her court, 1967.

Former Strawberry Festival queens were honored onstage at the 1959 festival.

DRUGS
CEN

- Chapter Two -

THE FLORIDA STRAWBERRY FESTIVAL

The Early Years

By Gil Gott

THE 1930S: THE FORMATIVE FIRST YEARS

It is traditional for agricultural societies to celebrate the fruit of their labors with a festival. As the strawberry grew in economic importance to the greater Plant City area, and with the looming shadow of the Great Depression, it became essential that the delightful strawberry has its own celebration.

The Plant City Board of Trade, the forerunner of the chamber of commerce, held a "Strawberry Day" as early as January 1, 1914, and residents were encouraged to promote Plant City's crops and tourism by sending postcards to distant friends and relatives.

1930

At a meeting of the Plant City Lions Club on July 5, 1929, Albert Schneider, the first president and organizer of the club, suggested the members sponsor a festival that would involve Plant City and the surrounding communities. The club members accepted the idea and formed a committee to start planning. The city quickly got on board and City Manager John C. Dickerson was named general manager of the nascent festival organization. The city then advanced $1,000 for the committee's operating revenue to be repaid from the festival's first proceeds.

The organization's incorporating charter was approved by the State of Florida and immediately, at a meeting at city hall, the strawberry festival organization elected its officers and directors: Albert Schneider, president; Walter Dee Marley, vice president; Henry H. Huff, secretary; Henry S. Moody, treasurer; and James W. Henderson, F. E. Cummins, and Marcus Cone, directors.

The newly incorporated Florida Strawberry Festival Inc. (January 6, 1930) drew up plans for the inaugural festival, which would include the selection and crowning of a festival queen. The site chosen was a vacant block owned by Ira M. Allen, a builder and developer who had homes in Michigan and Plant City. The site ran north from Baker Street between the Seaboard Railroad tracks and Michigan Avenue, just south and in view of Stonewall Jackson Elementary School. It was centrally located and over the years had been used for baseball and other carnival-like activities. It was Ira Allen who built the original festival building.

The dates chosen were Wednesday, March 12 through Saturday, March 15, 1930. The festival queen, who was to be selected through a contest sponsored by the Future Farmers of America, would be featured in the Grand Parade on opening day and would be crowned at the festival grounds.

In addition to the parade and the queen contest, the festival group developed plans for floats; vendors; a midway with rides; the Johnny J. Jones Exposition Shows; bands; demonstrations with plants, flowers, and a variety of food; and, of course, fresh strawberries. Johnny J. Jones, son of Welsh immigrants to Pennsylvania, was known to run a clean operation suitable for children, which continued to be a criterion for the festival.[1]

The story broke in the *Plant City Enterprise* on January 3, 1930, stating the dates and plans for the festival. The *Plant City Enterprise*, under Fenton M. Prewitt, and the *Courier*, under Wayne Thomas, both published every Tuesday and Friday and kept a running commentary on the festival's developments. The prize necklace the Future Farmers of America was to present to the new festival queen was on display at Edgar Hull Jewelry in the Wright Arcade Building.

By February 18, 1930, all the festival booths had been reserved; excitement filled the air and the suspense of the festival queen contest was prominent in the excitement. Plus, the Woman's Club of Plant City and the Home Demonstration Club had agreed to present their third annual flower show in conjunction with the festival. The communities of Dover, Hopewell, Seffner, Trapnell, Lithia, Turkey Creek, Cork, and Springhead had all registered to participate in the festival's activities.

Announcements were made in the two local newspapers that nominations for festival queen were open and that any "maid" or "matron" residing in east Hillsborough County was eligible. They could be nominated by clipping a form out of the newspaper, completing it, and submitting it to Gray Miley, who was working with the sponsoring Future Farmers of America Committee. The last day to nominate someone for queen was Monday, February 24. Voting would continue into March, lasting about two weeks.

Ballot boxes for votes for the queen were placed in stores about town to receive votes from the community. Each candidate would receive five thousand votes as sort of a starting bonus. After that, each vote was one penny, and pennies for votes were placed in clearly marked envelopes. Any resident of east Hillsborough could vote as many times as he or she wished—they just deposited the pennies with their votes.

The balloting for festival queen began in earnest on Tuesday, February 25. The *Plant City Enterprise* and the *Courier* would print the ten leading candidates on subsequent publication dates. The highest vote getter would be elected festival queen; the next five highest vote getters would make up the court. Heralds and pages were chosen separately. It is reported that Nettie Simmons, a girl from Dover, was in the lead early but was soon passed by Irvin Hopkins Wilder of Midway, the daughter of Calffrey LaFayette and Joanna Singletary Wilder. Irvin Wilder's lead gave way to Charlotte Rosenberg, daughter of Sam and Fannie Leibowitz Rosenberg, who had a retail business in Plant City.

Charlotte Rosenberg's vote totaled nearly fifty-thousand, raising almost $500, and she became the first Florida Strawberry Festival Queen.[2] Her court was comprised of Irvin Wilder, Nettie Simmons, Alice Sly, Kathryn Dudley, and Alice Maxey. Elected as heralds were Elizabeth Carey, Marian Herring, Elizabeth Hull, and Eugenia Sanchez; the pages elected were Katherine Andrews, Virginia Dennison, Catherine Fletcher, and Helen Spear.[3]

Thus, the newly elected queen and her court began to prepare themselves for the festival and the Grand Parade. The festival would begin Wednesday, March 12, with the Grand Parade stepping off at 1 p.m. The coronation

Charlotte Rosenberg, first Florida Strawberry Festival Queen, 1930.

THE 1930 QUEEN CONTESTANTS

By February 24, 1930, the last day to nominate a candidate for queen, there were thirty-six names submitted. Here are thirty-three of them.

Minnie Lee Alderman
Elizabeth Alsobrook
Mrs. Earl Barker
Mrs. Charles Daniels
Mrs. R. M. Fletcher
Ewana Glaros
Annie Ruth Harvey
Sarah Henderson
Mabel Hilsman
Mrs. Edgar Hull
Mrs. J. Edgar Knight Jr.
Mrs. Harry Lomison
Madie Lowry
Alice Maxey
Stella Moody
Anna Mae Nettles
Louise Nettles
Ruby Pratt
Charlotte Rosenberg
Mamie Sapp
Helen Simmons
Nettie Simmons
Alice Sly
Edna Smith
Margaret Strickland
Dorothy Sutton
Leola Sweat
Opal Clair Thomas
Hester White
Hazel Wiggins
Irvin Wilder
Daisy Woodall
Viola Yates

The entrance to the Florida Strawberry Festival at the east side of the fairgrounds, circa 1930.

of the queen was scheduled for 2:30 p.m. This was followed by a band concert, entertainment, vendors and rides, and more. The remaining festival days were Thursday, March 13—Strawberry Day; Friday, March 14—Tourists Day; and Saturday, March 15—Farmers Day.

On opening day, the queen and her court led the parade of elaborately decorated floats, decorated cars, horse-drawn carriages, and marchers. They started on Wheeler Street north of Risk/Herring Street and wound their way through the crowded streets in the historic district. The route consisted of Wheeler, Baker, Franklin, Reynolds, Evers, Haines, Collins, and Reynolds again, until turning onto Michigan and marching north to the festival grounds. The band struck up a march for the queen and her court as they promenaded to the throne.

The queen was preceded by her four heralds. Two pages bore her train and two more carried her crown and her necklace. The platform was decorated with a red, white, and blue bunting background and an overdrape of blue material with a silver fringe. Potted palms and fern enhanced the throne in the center of the stage. Albert Schneider, festival president, introduced Mayor-Commissioner George A. Carey, who spoke briefly before crowning the first queen.

The festival coincided with an annual event of the National Business and Professional Women's Club (BPW), and the local chapter decided to combine its observance with festival activities. On Thursday evening, March 13, the BPW hosted a banquet honoring festival officials and the queen and her court. The banquet was held at the Masonic Hall on Evers Street just north of Mahoney Street and was catered by women of the Eastern Star.

The official estimate of the attendance at the first day of the festival was 15,000, which was impressive in that the weather was inclement and not a chamber of commerce day. It was a great day for the officials, the queen, her court, the strawberry farmers, and the many volunteers who together produced the first strawberry festival.

Children enjoy the festival games and rides on the midway (left and center), circa 1930. Sideshows of every kind are also found on the midway at the 1930 festival (right).

1931

Joyfully reeling after their first highly successful festival, the organizers immediately began planning the next annual festival—setting the dates, critiquing the operations, lining up participants, selecting committees, etc. By December 1930, plans were well drawn and implementation had already begun. The 1931 festival, they believed, would be even bigger and better than the last.

Henry Hamilton Huff, Florida Strawberry Festival Association secretary, wrote his "Brief History Of the Beginning Florida Strawberry Festival" and had it printed in the new program. He wrote:

> The Lions Club, the youngest civic organization in Plant City had advanced a worthy idea and, not only had they simply advanced it, but they did not stop until all the other organizations had been set favorably towards it.
>
> Other organizations fell in line. The Kiwanis club, with their usual progressive methods, as well as the Plant City Woman's club – The Business and Professional Woman's club – the American Legion – the Beta Club – city officials, in fact, every citizen faced the issue with splendid co-operative effort.
>
> It would be impossible to bring out in detail the many problems that these citizens faced and met with a firm and aggressive attitude. However, the spirit shown by the citizenry as a whole was such as to urge them on and to those attending our Second Florida Strawberry Festival, we desire to say we are proud of our work and our efforts will continue, co-operating with other citizens to the end that the Florida Strawberry Festival will be made a permanent yearly exhibit – and to this we pledge our efforts – our enthusiasm – our time and our civic pride.

When the time came, Huff was right. The *Plant City Enterprise*, on Friday, March 6, 1931, carried this statement: "The Second Annual Florida Strawberry Festival, with two days to go, promises to prove an even greater success than last year's exposition." It continued, "The gates were thrown open Tuesday afternoon after a fine parade, approximately a mile in length, which was viewed by thousands of people who lined the streets along the route. The parade was acclaimed as the finest yet produced here."

Stepping off at 1:00 p.m., the parade was led by Plant City's own Battery E of the 116th Field Artillery, followed by scores of floats from Plant City, Tampa, St. Petersburg, Winter Haven, and Ybor City along with numerous decorated cars for clubs and officials. Bands from Plant City, Tampa, and Lakeland marched through the downtown district, and the parade traveled east on Baker Street to Michigan Avenue, then north

A float entertains the crowds at the 1931 Grand Parade through downtown Plant City.

to the exposition grounds for judging, after which the parade disbanded. The formal opening of the 1931 Florida Strawberry Festival followed at 2:00 on Tuesday, March 3. The theme was "School Day." A concert followed at 2:15.

Thirty minutes later, the introduction of guests and the speeches began; the master of ceremonies was popular District 4 County Commissioner W. T. Watkins. Other speakers included Albert Schneider, festival association president; Gerald H. Bates, mayor-commissioner of Plant City; County Commission Chairman W. T. Williams; George Carey, Festival Entertainment Committee chairman; Tampa Mayor Donald Brenham "D. B." McKay; Winter Haven Mayor O. P. Warren; W. G. Brorein, president of the South Florida Fair; J. P. Mays, president of the Florida Orange Festival; Carl Brorein of the Tampa Junior Chamber of Commerce (Jaycees); and Ira M. Allen, who, according to the *Plant City Enterprise*, "has generously aided the movement by erecting the buildings housing the Festival."

The Florida Strawberry Festival's Baby Parade queen and king are pictured in 1931.

Along with the berry packing contest and the Baby Parade, participants and visitors enjoyed the many displays, booths, and exhibits including strawberries, vegetables, citrus, canned products, and more. The fourth annual Woman's Club Plant City Flower Show, now a part of the festival and under the direction of the Woman's Club and the Home Demonstration Clubs of East Hillsborough County, was outstanding. It was the largest collection of flowers and plants ever viewed in Plant City.

The midway this year was put on by the Bernardi Exposition Shows and included "Valencia the Daring," the "Artistic Merrils," and the "Little Woman with the Giant Voice." The free acts and midway ran until midnight every night.

The second day of the festival was the "Crowning of the Queen of the Festival" scheduled for 2:00 p.m., with Mayor-Commissioner Bates doing the honors. The new queen was sixteen-year-old Irvin Wilder, a member of the prominent Wilder family. Her court included Bernice Adams, Gladys Balliett, Elizabeth Carey, Orel Ferguson, Elizabeth Hull, Genevieve McDermid, Elizabeth Morse, and Eleanor Murrill. The heralds and pages were Virginia Dennison, Jane Day Page, Ned Haven, and Reece Smith.

"Tourist Day" was celebrated Thursday, March 5, with an "Address of Welcome" by George A. Carey at 11:00 a.m. The free acts, exhibits, booths, and Bernardi Exposition Shows ran until

Sideshows along the midway at the Florida Strawberry Festival, circa 1930–1931.

midnight. A special event for the evening was the fiddler's contest held at 8:30 p.m.

Friday, March 6, was "Strawberry Day" and featured a general inspection of all displays by city officials and officers of the Strawberry Festival Association. That evening a dinner was held at the Hotel Plant for the queen, her court, public officials, and festival directors. It was a sumptuous affair.

The final day of the 1931 festival was "Farmers Day" and in addition to the shows, acts, rides, booths, and displays was an "Address to Farmers" delivered by former city commissioner and former county commissioner E. J. DeVane. The festival ran one day longer in 1931 than 1930, expanded from four to five days, and was experienced and enjoyed by thousands. The 1931 Florida Strawberry Festival succeeded in surpassing its own spectacular inaugural festival of 1930. It was now firmly established as a major annual event in Plant City and east Hillsborough County.

1932

With two strawberry festivals successfully completed, the organizers began to relax in their belief that the festival was now firmly established. Albert Schneider, now sixty-five years old and feeling the festival was on a sound basis, decided to retire. W. D. Marley, who had served as vice president for the first two years, stepped up to the presidency. James W. Henderson, who took over as general manager in 1931 after John C. Dickerson, also took office as vice president.[4] Henry Hamilton Huff continued as secretary and Henry S. Moody as treasurer.

The association also added a board of directors to its structure, adding George A. Carey and F. E. Cummins. The Executive Committee was formed, consisting of thirteen additional members, six of whom were females. They increased the committees from fifteen to seventeen and continued its separate panel of judges for exhibits and floats.

Secretary Huff concluded his "Brief History of the Beginning of the Florida Strawberry Festival," printed in the 1932 program, in much the same way as his 1931 history with these words: "To those attending our Third Florida Strawberry Festival, we desire to say we are proud of our work and our efforts will continue, co-operating with other citizens to the end that The Florida Strawberry Festival will be made a permanent yearly exhibit – and to this we pledge our efforts – our enthusiasm – our time and our civic pride."

The festival's 1932 queen, Elizabeth Carey, rides on her float in the 1932 Grand Parade.

Bear in mind that this was 1932, in the midst of the worst financial times America had ever experienced. Yet, Plant City area residents were surprisingly resilient, with positive outlooks. They had no idea what the future would bring, if the festival would really catch on or if the economy would soon take them all down. If nothing else, they were optimistic and tenacious.

And so, it began—the 1932 festival. The festival opened Tuesday, March 8, at

10:00 a.m. It was called "School Day." The Grand Parade began at 1:00 and wound through the streets of Plant City arriving at the festival grounds for a formal opening with rocket salutes at about 1:30. Festival President W. D. Marley gave the welcoming address at 2:30, followed by Mayor-Commissioner George A. Carey, who also chaired the Entertainment Committee.

At 3:00 the "Free Acts at the Grandstand" started up, featuring the "World's Greatest High-Wire Acts" with DoBell and party. Frederick DoBell was a sub-contractor signed on with Johnny J. Jones and provided an "Electro-Pyrotechnic High-Wire Spectacle!" DoBell was called the electrical wizard of the high wire and was named "the greatest free attraction ever seen in the west." He was followed by the Kuhn string trio and Madame Lascelle and Madame Kuhn, both of whom did vocal solos.

At 4:00 p.m. one of the local favorites, the berry packing contest, took center stage. The free acts were repeated at 8:00 and the Johnny J. Jones World's Greatest Midway Attractions continued until midnight. In addition to the rides, the midway in those years was a carnival, with the usual freak shows, magic shows, and sometimes a little burlesque flavor.

Wednesday, March 9, was "Governor's Day and Crowning of Festival Queen" and the gates opened at 9:00 a.m. At noon the festivities took place at the downtown Hotel Plant, where the festival directors honored Governor Doyle Elam Carlton and his wife, Nell. Governor Carlton was comfortable in Plant City and with farmers; he was a native of Wauchula and although he was a Columbia University-educated attorney, he was involved in agriculture throughout his life.

Along with the festival directors, Governor Carlton inspected the exhibits at 2:00 p.m. and gave his address at 3:15. It was a busy day. After the four o'clock berry packing contest and the eight o'clock repeat of the free acts came the finale. After introductions at 8:30, Governor Carlton proudly crowned Elizabeth Carey, seventeen-year-old daughter of Mr. and Mrs. George A. Carey, as the 1932 Florida Strawberry Festival Queen. The stage was festooned with potted plants and flowers and the queen and her court, pages, and heralds were bedecked in colorful costumes. It was an elegant and exciting moment. Plant City Mayor George A. Carey and Governor Carlton were close friends, and in 1931 the governor had appointed Carey to the Hillsborough County Budget Commission. They enjoyed this crowning ceremony.

Thursday, March 10, was "Agricultural Day" and began with the opening of the gates at 9:00 a.m. The noon luncheon at the Hotel Plant was in honor of Nathan Mayo, who kept his home and farm in Summerfield, Marion County, and who had served as the commissioner of agriculture since 1923. Mayo inspected the exhibits at 2:00 p.m. and gave his address at 3:45, following the performances of DoBell, Mesdames Lascelle and Kuhn, and the Kuhn trio. (Mayo would go on to become the longest serving commissioner of agriculture in Florida history, serving thirty-seven years and dying in office in 1960. He is credited for establishing the State Marketing Board, which

Florida Strawberry Festival Queen Elizabeth and her court at the coronation ceremony, 1932.

led to the construction of State Farmers Markets, of which the largest was constructed in Plant City in 1939.)

After the strawberry packing contest and the free acts, the evening closed with a spectacular firework display at 10:00 p.m.

"Hillsborough County Home-Coming Day and Latin Day" was the title for Friday, March 11. After the gate opening at 9:00 a.m., the midway attractions, exhibits, and displays were in full gear. The luncheon at the Hotel Plant was in honor of Robert E. Lee Chancey, recently elected mayor of Tampa. Chancey completed his inspection of the exhibits then made his address and comments at 2:00 p.m.

Another developing favorite event at the festival was the Baby Parade. At four o'clock came the "Grand Parade Baby Floats and Beauty Contest, ending at the Grandstand." This took place under the watchful eye and guidance of Miss Jane Forehand, well-known Plant City kindergarten teacher. It was a beautiful, fun, and exciting parade with costumed children, imaginative floats, and nervous parents. The first round of the strawberry packing contest followed at 5:00 p.m. and the delightful fireworks display again signaled the end of the evening's program.

The last day of the 1932 Florida Strawberry Festival was on Saturday, March 12, and it was called "Everybody's Day." The gates opened again at 9:00 a.m. and the honored guest at the noon luncheon at the Hotel Plant was the Honorable Pat Whitaker, state senator from Tampa. Whitaker was a flamboyant attorney, was very articulate, and was the president of the state senate from 1931 to 1933. He was also the brother-in-law of Robert E. Lee Chancey, mayor of Tampa. He gave his entertaining address at 3:45 p.m.

The second round of the strawberry packing contest was held at 4:30, with the contest finals being held at 5:30 at which time the trophy winners would be announced. Then, at ten o'clock was the first of the American Legion-sponsored automobile raffles at which a brand new Chevrolet would be given away at the grandstand! At midnight, the 1932 Florida Strawberry Festival ended with the words: "Good – Bye, Good Luck and Lots of Prosperity."

1933

With the impact of the Great Depression worsening, 1933 quickly became an inauspicious year. As the economy continued to falter, President Herbert Clark Hoover became more and more unpopular. In November 1932, he lost his re-election campaign to Franklin Delano Roosevelt, who promised a New Deal.

It was an eventful time. By February of 1933 the calamitous economic effect of the years of prohibition had finally resulted in congressional action, and the Blaine Act to end prohibition was passed on February 17. Despite an attempted assassination, in which President-elect Franklin Delano Roosevelt was not injured, the country began to hope there was a light at the end of the tunnel. Roosevelt was inaugurated on March 4, 1933, and immediately swung into action establishing numerous programs under the rubric "New Deal."

For the festival, things were challenging at first. Ira Allen, whose land the festival had been using and who built the first buildings for the festival, died in 1932 at the age of seventy-six. His daughter Elsie, then about thirty-eight years old, stepped in to continue her father's work in real estate development, etc., and continued to support the festival. The festival continued to use the same area for their fairgrounds for a few more years.

Following three successful festivals in 1930, 1931, and 1932, the planning for 1933 went smoothly. Few changes were made, amongst those were adding City Commissioner George Carey as vice president, along with current Vice President and General Manager James W. Henderson, and adding Plant City Mayor Gerald Bates to the board of directors.

Among the returning Executive Committee members was Percy Page, one of the original Lions Club members promoting the concept of a strawberry festival, who again this year served as chairman and marshal of the Grand Parade. Page was a World War I veteran and a member of the Lions and the American Legion; his daughter Jane Day Page was a herald for the 1931 Florida Strawberry Festival Queen.

As in 1932, the first day of the festival, Tuesday, February 28, was Grand Parade Day. The gates opened at 10 a.m. and the parade stepped off at 1 p.m. The marchers, bands, floats, and bedecked cars with honorees paraded through the city streets ending up on the festival site on Michigan Avenue. Beginning at 1:30, festival officials George Carey, vice president; W. D. Marley, president; and Mayor Gerald Bates all gave welcoming addresses expressing the goodwill of the festival organization to the guests.

The "Free Acts" came on at 3 p.m., repeated at 7:30, and included "The Aerial Bauers, in high trapeze acts," "The Happy Sisters" in rollicking songs, and the Matsumoda Troupe in balancing and slack-wire acts and featuring Matsumoda's "Slide for Life" along with "Buck" Buchanon in lightning artistic sign painting. From opening until midnight, the Model Shows of America provided illuminated midway attractions—shows that would interest and mystify! The evening wrapped up with the "awarding of prizes at the Grandstand," although the program does not mention which prizes.

The Business and Professional Women's Club's ornate float makes its way along the parade route in 1933.

Again, the second day of the festival was Queen's Day—the crowning of the queen. The gates opened at 9:00 a.m. with special features and announcements at the grandstand. The free acts performed at 3 and 7 p.m. followed by the crowning event unveiled at 8:30. Queen Christine Walden was crowned in an elaborate ceremony under the direction of the Parent-Teachers' Associations, supervised by Colonel J. Edwin Cassels. Her court included maids Christine McDonald, Lillian Schulte, Mary May, Mildred Calding, Dorothy Adelson, Nona Mae Holloway, Anne Vannerson, and Eloise Horton; crown bearer Jack Fewell; heralds Gloria Patterson and Marion Miles; and pages Walter C. Busk Jr. and George Carlton. The evening concluded with the "Prize Award at Grandstand," and the midway's Model Shows of America continued until midnight.

Keeping true to their pattern, the festival directors honored an outstanding community leader on Thursday, March 2. It was Brorein Day, in honor of William Gebhard Brorein, founder of Peninsular Telephone Company, civic leader, member of the Knights of Pythias, president of the Tampa Board of Trade, and president of the South Florida Fair Association (later known as the Florida State Fair Association). Brorein was treated to a luncheon at the Hotel Plant by the festival directors and city officials.

At 2:00 p.m. Brorein inspected the exhibits along with the festival directors and at 2:30 the ceremonies at the grandstand began with James W. Henderson, festival vice president and general manager, introducing Brorein, who proceeded to address the crowd. The free acts ran at 3:30 and again at 7:30 p.m. At 10:30 was the awarding of prizes at the grandstand and the midway was open until midnight. Missing from this year's activities were the strawberry packing contest and the evening's closing fireworks.

The souvenir program for the fourth annual Florida Strawberry Festival in 1933.

Friday, March 3, was both Agricultural Day and the annual Parade of Babies. The gates opened at 9:00 a.m. with morning grandstand features and announcements. Being Agricultural Day, the festival honored both Nathan Mayo, commissioner of agriculture, and L. M. Rhodes, commissioner of the State Marketing Board. They toured the exhibits at 2:00 p.m. and Mayo gave his address at 2:30. The free acts followed at 3:00 and again at 7:30.

The highlight of the day was the Baby Parade at 4:30 p.m. The children, kindergarten age or younger, paraded by in their glamorously decorated floats, wagons, and more. Some dressed like Liberace or Lou Gehrig, some with their pet animals, to be judged on beauty, health, and personality. Kindergarten teacher Jane Forehand orchestrated the exciting event.

Concluding the honors that evening was Commissioner L. M. Rhodes, who gave his address at 7:00 p.m., followed by free acts and the midway attractions until closing at midnight.

Running five days in those early years, the final day of the festival was Saturday, March 4, the same day as the inauguration of President Franklin Roosevelt. This was called "Everybody's Day" and consisted mainly of free acts and midway attractions. The American Legion ceremonies and automobile raffle took place at 10:30 p.m. and the fair officially closed with a "Good-Night, Good-Bye and Good-Luck" at midnight.

1934

The 1934 festival approached with most of the organizers feeling confident that a high-quality festival was becoming something to be expected of them and that they would meet the challenge. The world about them, however, was still far from being stable.

President Roosevelt had been in office almost a year, and his many New Deal programs had only begun to have a positive effect. There were still millions unemployed (22 percent), and the Dust Bowl continued to deliver its disastrous blows to the Midwest. For many, the outlook was bleak. Other events in 1933–1934 included Germany refugee Albert Einstein visiting the White House; the Twenty-first Amendment, repealing prohibition was finally passed; and Frank Capra's movie *It Happened One Night* with Clark Gable and Claudette Colbert opened to crowds in New York. Gasoline was ten cents a gallon, and the median monthly rent was $20. Average annual income was stuck at $1,368.

In Plant City, there were only a few changes to the association in preparing for the 1934 festival and attempting to add more volunteer

Florida Strawberry Festival Queen Dorothy Adelson and her court, 1934.

power to the organization. They doubled the size of the board of directors from seven to fourteen, adding Gerald Bates, Arthur Boring, F. D. Goff, C. G. Green, L. T. Langford, Percy Page, T. E. Rogers, and William Reece Smith. Treasurer Henry S. Moody stepped down and Henry H. Huff became both executive secretary and treasurer. Gerald Bates was elected vice president, and James W. Henderson was elected vice president general manager.

Again, in preparation for the 1934 festival, Plant City participated in Tampa's South Florida Fair, which had changed its name to the Florida Fair. That fair was the outgrowth of promotions begun by Henry Plant to bring patrons to his Tampa Bay Hotel. The Plant City Community Booth won first honors for community booths for the third year in a row.

The big kick-off for the 1934 festival was the Grand Parade, leaving its staging area at 1:30 p.m. and wound its way through the city streets to the fairgrounds where the fifth annual Florida Strawberry Festival was officially opened. Festival association dignitaries welcomed the crowd. A new and greater variety of free acts, including one called "Eight Dogs and Two Cats," entertained the festival guests, while the Royal Palm Amusement Company's midway shows continued to run until midnight.

The queen pageant had not yet been established, and the queen selection was still determined by the penny-a-vote method begun in 1930. But the coronation itself was a lavish and spectacular event. This year the coronation was preceded by the earlier official visit of the Governor and Mrs. David Sholtz, who attended a luncheon and an evening reception sponsored by the Woman's Club and the Business and Professional Women's Club at the new library. Later they were honored at a dinner at the Hotel Plant, where they were joined by festival officials and their wives.

The actual crowning of the new Strawberry Festival Queen, Miss Dorothy Adelson, lovely daughter of Mr. and Mrs. Samuel Adelson, took place that Wednesday evening, beginning at 8:30. It was an outstanding event with the flourish of gowns, flowers, pages and heralds, and the queen with her shining tiara. Fittingly, in years to come, this event would take place at Adelson Field, named in honor of Samuel Adelson.

Thursday, March 1, was Agricultural Day at the festival, and once again Commissioner of Agriculture Nathan Mayo was honored at a luncheon at the Hotel Plant and was the main speaker in the afternoon program at the fairgrounds. Mayo had an agricultural background and enjoyed his visits to the Florida Strawberry Festival.

In the program for the festival that year, one full page was devoted to the Plant City Flower Show, noting that it began prior to the Florida Strawberry Festival and after two successful years was then held in conjunction with the festival in 1930 through 1933. This year the show was to be bigger and better than ever. The flower show was "one of the drawing cards of the Festival." The exhibits were open until 10:00 p.m. every night of the festival.

Friday, March 2, was Strawberry Day and Grand Baby Parade and the beauty, health, personality contest. The babies and their floats set off at 4:00 p.m. in a short parade which ended at the grandstand, where both the babies and the floats were judged. Saturday was Hillsborough County Home-Coming Day and featured basically the free acts and the midway shows. "Special invited guests were to take part in an impromptu

The always entertaining Baby Parade in 1934.

program to be arranged after arrival." Everything came to an end with a "Good-Night, Good-Bye and Good-Luck."

On the program page listing the fifty-two booths in the exhibit buildings, some stood out. Home demonstration agents had participated in organizing events for the festival from the inception in 1930, and this year set up booths in the exhibit buildings. There were two: the Home Demonstration Exhibit was under "Miss Ruth" and the Colored Home Demonstration Exhibit was under Floy Britt. A graduate of Florida A&M, Miss Britt joined the Florida Cooperative Extension Service in 1932, later rising to the office of state director of 4-H clubs for African American youth. She visited Bealsville and the Plant City area frequently.

There were also three booths for the Future Farmers of Florida, but none for the Future Farmers of America. FFF was prominent in Florida for a number of years but was later discontinued.

1935

Although the mid-1930s began to show some lifting of the hope of recovery, poverty was still widespread. The national economy continued in its decline as the Great Depression dragged on. Fortunately, like communities throughout America, Plant City benefitted from New Deal programs. The projects developed under the Civilian Conservation Corps (CCC) and the Works Progress Administration (WPA) kept many residents of the Plant City area working.

The 1935 Florida Strawberry Festival was set for Tuesday, March 5 through Saturday, March 9—five full days of festivities, exhibits, rides, shows, parades, and speeches. The Grand Parade started at 1:30 p.m., ending near the fairgrounds for the 2:30 formal opening where the crowd was welcomed by the festival dignitaries and the mayor. The free acts began early and the Royal Palm Shows Midway entertained the visitors until closing at midnight.

Wednesday, March 6, was Governor's Day and the coronation. Governor David Sholtz returned to visit the festival again and was the guest at the noon luncheon at the Hotel Plant. He delivered his address at 3:45 p.m. at the grandstand. The crowning of the queen, Miss Virginia Moody, began at 8:30 p.m. and again featured an array of gowns, flowers, and heraldry.

Like festivals in the past, this year the honored guest on Thursday was Peter O. Knight of Tampa. Knight was a respected lawyer, prominent businessman, president of Tampa Electric Company, and in 1934 received the Civitan Club Award for his public service. He was the honored guest at the Hotel Plant luncheon and, introduced by Mayor Gerald Bates at the 4:00 p.m. program, Knight was the main speaker.

Friday, March 8, was Strawberry Day and the Grand Baby Parade. Once again, the honored guest

Plant City Mayor Gerald H. Bates pays tribute to 1935 Florida Strawberry Festival Queen Virginia Moody.

Queen Virginia Moody and her court at the 1935 coronation ceremony.

at the luncheon was Agriculture Commissioner Nathan Mayo. He gave his address at 3:30 p.m. followed at 4:00 by the Grand Baby Parade of toddlers and floats under the direction of kindergarten teacher Jane Forehand. As usual the parade was stunning, with highly imaginative floats and costumes—and nervous parents.

The last day of the 1935 festival was American Legion Day and featured Plant City's own Arthur R. Boring, president of the chamber of commerce and past commander of the Florida American Legion. Additional speeches were made by other statewide prominent Legionnaires. And the Royal Palm Amusement Company midway shows and attractions closed at midnight to conclude another wonderful Florida Strawberry Festival.

The Baby Parade always draws a crowd, 1935.

Another glimpse of the popular Baby Parade shown here in downtown Plant City in 1935.

1936

In office since March of 1933, by 1936 President Franklin D. Roosevelt and his New Deal plans were beginning to have positive effects on society and the economy. The WPA and the CCC put Americans back to work in public programs, some of which benefitted Plant City very well.

On the down side, however, was a major social problem in America—poverty among the elderly. Although the Social Security Act was passed in 1935, the Social Security Administration had yet to be established. To deal with the problem of old-age poverty, in 1933 Dr. Francis E. Townsend issued his own plan to help the elderly. It became immediately popular and the demand for $200 per month for the elderly grew rapidly.

Across the nation, groups supporting the Townsend Plan were formed and promoted the concept. President Roosevelt objected to it because it did not require any employee payments, and Congress passed the Social Security Act as a means of quieting the populous movement. It was a first step towards what became a full-blown government program over the next several decades. Nevertheless, Townsend groups throughout the nation continued to promote the Townsend Plan for years.

So it was that in 1936 the Florida Strawberry Festival held its Townsend Day on Thursday, March 5. Festival goers would hear a "prominent Townsendite" extoll the virtues of the plan to help the impoverished elderly among the citizens.

The officers of the festival association in 1936 were the same as before—the experienced and competent George Carey served as president, and James W. Henderson ("Coca-Cola Jim") continued to serve as vice president and general manager. Once again, the Plant City community participated in the February Florida Fair in Tampa, along with the Gasparilla Parade. The float for the queen and her court had its first trial in that parade and was usually judged well. The booth in the community booth competition won first honors again this year.

The Florida Strawberry Festival Community Booth took first place at the South Florida Fair in Tampa in 1936.

Tuesday, March 3, was Grand Parade Day in Plant City and Miss Virginia Moody, the reigning festival queen, rode on the bedecked float as the parade stepped off at 1:30 p.m. The parading groups and celebrities crossed the finish line just west of the SAL tracks. They then headed for the opening of the 1936 festival at 2:30.

Wednesday, March 4, was Governor's Day and the crowning of the festival queen, with Governor Dave Sholtz and the 1935 and 1936 festival queens being honored at the Hotel Plant luncheon. (Details why the crowning was postponed are unknown, but the coronation planned for Wednesday night did not occur until Friday.) After the governor's address at 3:45 p.m., the visitors enjoyed the free acts and the American Model Shows midway until midnight.

Thursday was Townsend Day and Plant City Commission Men's Day. At 2:30 p.m. the Townsendite speaker addressed the crowd at the grandstand promoting the Townsend Plan, and at 7:30 that evening the ceremonies were in honor of Plant City's Commission Men and prominent businessmen.

Plant City Mayor Gerald H. Bates proudly crowns Pauline Schulte as the 1936 Florida Strawberry Festival Queen.

Friday was a busy day for the festival. It was Farmers' Day and the Baby Parade was held. The honored guest at the luncheon was Agriculture Commissioner Nathan Mayo, who, after lunch, addressed the crowd at 1:30 p.m. The Baby Parade began sharply at 2:30 and the costumed youngsters paraded by on foot or on their floats for the judging on the attributes of "Beauty, Health, and Personality." The judging also included the originality and integrity of the floats. Friday night, according to the Tuesday, March 10, *Courier* article, the coronation went well: "A silver moon, peering down from behind a veil of restless clouds, last Friday night saw Queen Virginia of the House of Moody step down from the royal throne of this winter strawberry kingdom and Queen-elect Pauline of the House of Schulte ascend that throne midst a colorful scene of pageantry and enticing music. While the moon peered and saw, several thousand persons also acclaimed the new monarch, paid homage to the departing Queen."

The final day of the 1936 festival was Florida American Legion Day. Although the Lions Club had originated the concept of the festival, it was becoming clear the American Legion had become a major supporter of the event and provided new life and direction. Plant City's own Arthur R. Boring served as state commander of the Florida American Legion and was the honoree of the program. He brought a number of American Legion dignitaries with him. It was a fitting program for the closing of the seventh Florida Strawberry Festival.

1937

As the Great Depression years wore on, the economy began to recover, unemployment was declining, and hope seemed just around the corner. Then, in 1937, the new recession began. Roosevelt had just begun his second term and, although things began to look better, there were many that still struggled.

In Plant City, the people were still optimistic and the plans for the eighth annual Florida Strawberry Festival were being finalized. In February, not only did Plant City win first honors for its community booth at the Florida Fair in Tampa, but the Florida Strawberry Festival float won first prize in the Gasparilla Parade—with 1936 Festival Queen Pauline Schulte and her court lovingly adorning the glamorous float.

The contest to elect a new queen heated up the minute the nominations were announced by the Nominating Committee shortly after opening the nomination ballot box. Twenty-three "charming Plant City and East Hillsborough Young ladies yesterday stood on the starting tape in the race for Queen of the eighth annual Florida Strawberry Festival." The contest was still decided by the penny-a-vote system initiated in 1930, and on the first official count of votes Miss Ann Eberhardt was in first position followed narrowly by Miss Catherine Fletcher. The remaining sixteen candidates were close behind, and five contestants had already withdrawn.

At six o'clock in the evening, the Monday before the festival would begin, the ballot boxes were gathered and opened by the committee. Miss Eloise Howell, descendant of the Pioneer Springhead family, led the voting and was announced as the queen-elect of the 1937 festival. The court included both Ann Eberhardt and Catherine Fletcher, along with Lauralou Harold, Marguerite Yates, Lucille McClellan, Virginia Dennison, Robbie Jarvis, and Marceil Booth. The lovely Miss Howell was to be crowned at a new coronation pageant on Thursday night, March 4.

The 1937 festival began with the usual Grand Parade through the downtown streets

WINNER OF FIRST PRIZE

The Florida Strawberry Festival float in the 1937 Gasparilla Parade in which Queen Pauline Schulte and her court rode to first prize for out of town entries. The prize was a beautiful silver loving cup.

The Florida Strawberry Festival's prize-winning float in the 1937 Gasparilla Parade in Tampa, Florida.

beginning at 1:30 p.m. and disbanding in time for the participants to attend the formal opening at 2:30. This time, after the welcoming comments of the festival dignitaries, Festival Association Vice President and General Manager Jim Henderson presented his review of the past festivals since its beginning in 1930. The visitors were then treated to the free acts and the Royal Palm Shows and Midway, open until midnight.

Wednesday, March 3, was Governor's Day and the honoree was the newly elected governor, Fred P. Cone, who at sixty-five years old and being a strident segregationist was not a favorite of the festival organizers. He gave a short address at 3:35 p.m., after which it was free acts and midway time.

No special events were scheduled for Thursday, March 4, other than the coronation ceremonies, which the new "pageant director" stated were to be royal and regal. The ceremony featured the queen on a glittering throne, her eight maids, four train bearers, two heralds, two flower girls, and one crown bearer. It was a beautiful sight.

Tourists' Day was the theme for Friday, March 5, but the main feature was the Baby Parade and the beauty, health, personality contest that began at 3:15 p.m. This year the event was under the direction of Dorothy Taylor Boring, the wife of festival official, former mayor, and city commissioner, Arthur Rice Boring. Mrs. Boring replaced the initial Baby Parade director Miss Jane Forehand.

The festival closed on Saturday, March 6, which was American Legion and Agriculturists' Day, and featured a luncheon at the Hotel Plant in honor of prominent Legionnaires along with the speaker of the day, the inimitable Agriculture Commissioner Nathan Mayo, who was enjoying becoming a regular at the festival. After the afternoon festivities at 3:30 p.m., with prominent visitors and a speech by Commissioner Mayo, the day closed with the free acts and the midway, which visitors enjoyed until midnight.

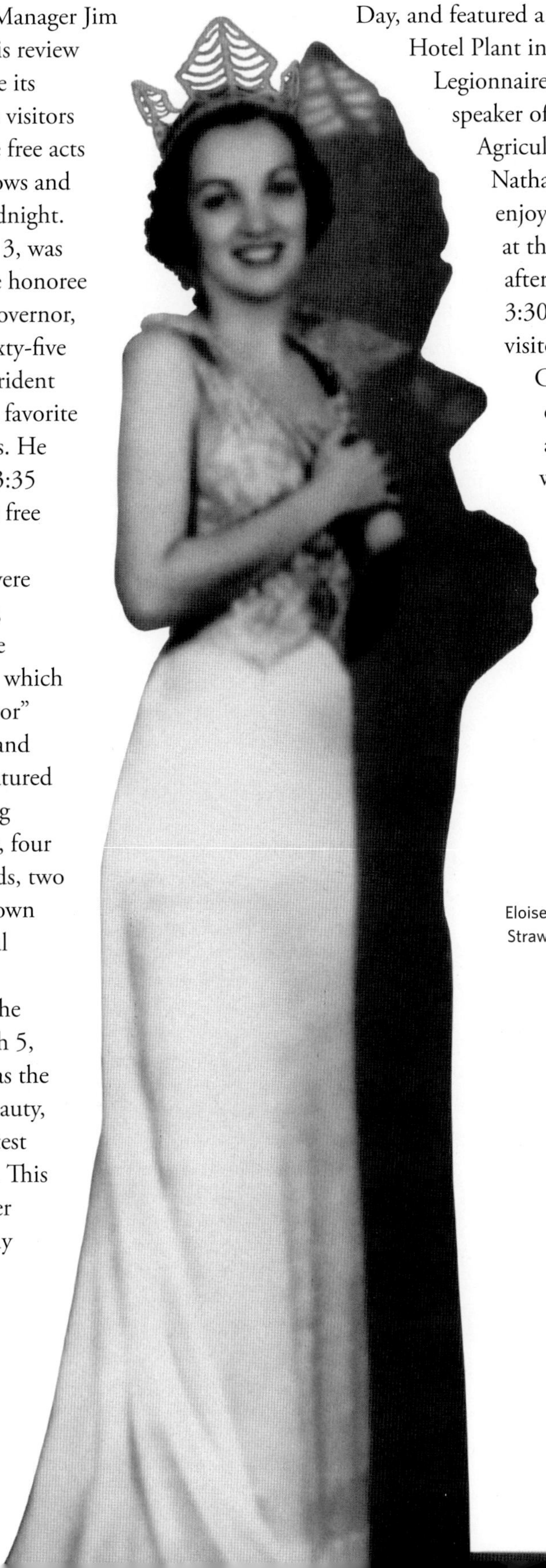

Eloise Howell was crowned the Florida Strawberry Festival Queen in 1937.

1938

An interesting article appeared in a Tampa newspaper in early February 1938 stating that the Florida State Fair in Tampa had become the "try out city for virtually all new inventions in ride devices and tent theatre attractions." The original "Flying Trapeze," overhead illumination, numerous high wire acts, and much more were tried first at the Florida State Fair before being taken on the road by the midway and carnival companies. Many of these same shows found their way to Plant City's Florida Strawberry Festival, which followed soon after the fair in Tampa. The popular Royal Shows company, with aerial shows and daredevil drivers, signed on for the 1938 festival.

Running from Tuesday, February 22 through Saturday, February 26, the 1938 festival was the earliest start to date of the five-day event. For 1938 the Florida Strawberry Festival was organized by fourteen members of the board of directors and the members of twenty-seven committees, including the coronation pageant, now under the direction of the Grammar Schools' Parent Teacher Association.

Opening day of the festival on Tuesday, February 22, focused on the Grand Parade, with floats, marching bands, and cars of celebrities and dignitaries, and was followed by the official opening of the festival and the welcoming by the festival association and city officials. Wednesday, February 23, honored Florida cabinet members and agriculture. The primary speaker was one of Plant City's favorites, Agriculture Commissioner Nathan Mayo.

There was no Governor's Day in 1938 and on Thursday, February 24, the featured events were a Grand Parade presented by the Dramatic Order of the Knights of Khorassan (DOKK), which was a division of the Knights of Pythias and specialized in comic antics, stunts, and fun-filled parades. This was the first time the DOKK (also called Dokkies) and the Knights of Pythias were in the festival events. Also that evening was the spectacular and highly decorative coronation with the queen, her court, heralds, pages, and all the trappings of

Cupid watches the youthful couple on this float in the 1938 Baby Parade.

royalty as Mayor J. Arden Mays crowned the lovely Miss Norma Robinson as the festival's 1938 queen.

Tourist Day and Baby Parade Day excited the people on Friday, February 25, and at 3:15 p.m. baby floats, youngsters marching, and others paraded through the streets to the festival grandstand for the final judging. Closing the festival, on Saturday, February 26, was Farmers' Day and American Legion Day, and the major focus was on the prominent state and county agriculturists who would speak about the advantages to be derived from the Plant City Farmers Market, a $247,000 WPA project currently under construction. The festival closed with this highly optimistic future.

Here are some of the youthful contestants in the 1938 Baby Parade.

With all the excitement, ceremonies, and speakers, so much more goes on at the festival, including the glamorous Woman's Club Flower Show. In 1938 there were also more than nine categories of fruits and vegetables, commercial and non-commercial, judged on canning, packing, displays, etc., taking place under the watchful eyes of committees and judges. There were also more than fifty exhibit booths, including the Colored Farmers' Agricultural Display under Elliot Robbins, the Colored Home Demonstration Booth under Floy Britt, the Frances Wordehoff Sunday School Class Lunch Room, the Coca-Cola Bottling Company booth, and, of course, a strawberry shortcake booth. It was a very successful ninth annual festival.

1939

Celebrating its tenth annual festival, the organizers relished their successes and were delighted to have their celebration along with the dedication of the new State Farmers Market at Plant City. The association adjusted its committees, hired the Silver Blossom Shows for its new midway attractions, and invited the DOKK to put on its comedy parade. The big night would be Thursday, March 9, as they held the ceremonial dedication of the State Farmers Market on site, then, with their usual grand style, held the tenth coronation ceremony at the festival grandstand.

All things considered, this was an admirable achievement—especially after having gone through the worst years of the Great Depression. The participants in the Grand Parade stepped off proudly at 1:30 p.m. to start the last parade of the 1930s. The formal opening of the Florida Strawberry Festival was at 2:30, with the festival association and city officials welcoming the visitors. Jim Henderson, the festival's general manager, also gave his welcome and a recap of their first decade.

The Wednesday program on March 8 was DOKK and Pythian Day and featured Dokkie officials, the Dokkie comic parade, and Dokkie ceremonies at the grandstand in the afternoon.

Thursday, March 9, was a very special day and the lunch at the Hotel Plant honored the State Farmers Market officials, welcomed by Dr. Calvin T. Young, chairman of the local State Farmers Market Board of Directors. In the evening the activity moved to the site of the new market near the intersection of Haines and Alexander Streets for the actual dedication ceremony led by Plant City's Ben Rawlins and W. L. Wilson, director of Florida State Markets. For the local farmers, it was a grand event and the promise of a better future.

Not to be outdone, at 8:00 p.m. was the grand coronation ceremonies back at the festival grandstand. With pomp and ceremony, surrounded by seven heralds, pages, train bearers, and eight courtly maids, reigning Queen Norma Robinson yielded her tiara to Mayor J. Arden Mays, who crowned the lovely Miss Faye Mott as the 1939 Florida Strawberry Festival Queen.

Friday, March 10, was called Tourist Day, but it was the Baby Parade that always stole the show. All day the parents and families were preparing their young entrants for this grand event, and at 3:15 p.m. they set off, parading through the city streets to the festival grounds, then onto the stage where the prizes were then awarded. There were many entrants and there were also many prizes—various categories and age groupings allowed for a wide distribution of loving cups and many ribbons.

The final day of the 1939 festival was Saturday, March 11, and the focus was both prominent American Legionnaires and state and county officials—but mostly the Legionnaires. The speeches were at 2:30 p.m. and the big event of the evening was the American Legion automobile raffle which was announced at a ten o'clock ceremony. This concluded the end of the first decade of the Florida Strawberry Festivals.

A postcard of the new Plant City State Farmers Market—the largest in the state of Florida in 1939. It was dedicated at a Florida Strawberry Festival special event.

THE 1940S: A NEW DECADE BEGINS, DIMS, AND IS REKINDLED

After ten years of producing successful festivals during the 1930s, the organizers met some challenges in the 1940s. Perhaps the first, and the most significant challenge, was the change of location. The Florida Strawberry Festival moved from the east side of Plant City, where it had been since its inception, to the west side where it could use the recently renovated Adelson Field. It was also adjacent to the American Legion facility, the Plant City National Guard Armory complex, and near the new (1939) State Farmers Market.

The parade route changed, reversing its direction, forming east of the Seaboard tracks near the Old Berry Yard and moving west on Reynolds, Collins, Haines, Evers, back to Reynolds, and then west to the "Fair Grounds at Ball Park and the grand parade participants would disband West of Festival building." Acts and events were staged around Adelson Field and adjacent areas. This was the precursor of the growth and development that was yet to come.

Another major change for the festival was the charge for admission. It varied for general entrance to the festival grounds (ten cents), for the queen coronation (twenty-five cents), or the midway (usually free). Sometimes ten cents or twenty-five cents were charged for other special events. Some things continued—the Silver Blossom Midway attractions returned and the Dokkies would again march in their comic parade.

1940

Among the forty-eight exhibit booths were the usual community booths—Pasco, Plant City, Dover, Springhead, and Turkey Creek—and the 4-H and Future Farmers, the Home Demonstration Exhibit, and the Colored Home Demonstration booth under Floy Britt. There were also the Frances Wordehoff Sunday School Class Lunch Room and the American Legion. This year also featured (for the first time) the State Marketing Board booth, of great interest to local agriculturists.

Having enjoyed the comedy and antics of the Dokkies parade in 1939, the Knights of Pythias were invited again and held its exciting comedy parade at 1:30 p.m. on Wednesday, March 6, marching and rollicking all the way from East Reynolds Street west to the new festival building at the relocated fairgrounds. This was the most notable event of the day.

Governor Cone, in his last year in office, was honored at the noon luncheon at the Hotel Plant. He gave his speech at 3 p.m. at the Adelson Field grandstand.

Perhaps the most important day of the 1940 festival was on Thursday, March 7. After holding its festivals from 1930 to 1939 at a leased lot on Michigan Avenue, the Florida Strawberry Festival Association had moved to a new and larger complex on West Reynolds Street. There they constructed their own building and fairgrounds and gained access to the American Legion building

Florida Strawberry Festival Queen Catherine Fletcher is shown here with her court, 1940.

The queen contestants ride enthusiastically in the 1940 Grand Parade.

and area, as well as Adelson Field and the National Guard Armory complex.

The dedication of the new Florida Strawberry Festival Fairgrounds was held at the Adelson Field grandstand at 2:30 p.m. that Thursday. The other special event was the spectacular coronation ceremony with all its royal trimmings. At the eight o'clock ceremony, the first in their new location, Miss Catherine Fletcher was crowned the 1940 queen by the Honorable Don Walden, mayor of Plant City. Ironically, Miss Fletcher's parents owned Fletcher Ford and the mayor owned Walden Chevrolet.

One of the festival's loyal friends was the Honorable Nathan Mayo, Florida's commissioner of agriculture, who was honored at the Friday, March 8, noon luncheon at the Hotel Plant. Joining Mayo were other state, county, Tampa, and Plant City officials. The commissioner spoke with great pride about the Plant City State Farmers Market, which was the largest of its kind in Florida.

But all the excitement that day was reserved for the 3:15 p.m. Baby Parade, for the first time being held at Adelson Field. Bubbling and enthusiastic youth paraded in front of the judges and nervous parents. In the end, ten loving cups and twenty ribbons would be awarded in two age groups: six months to two years and two years to four years. They were judged for best decorated float, most original float, most beautiful baby, healthiest baby, and personality. Mrs. Arthur (Dorothy) Boring supervised the wonderful event.

The final day of the 1940 festival came on Saturday, March 9, and few events were scheduled. The 2:30 p.m. "Ceremonies at the Grandstand" were labeled "Impromptu," and admission to the park was free from the opening at 9:00 a.m. until 7:30 p.m. when the fee was increased to ten cents. The big event of the final day was the 10:00 p.m. American Legion automobile raffle, after which the shows and midway attractions continued until midnight.

FLORIDA STRAWBERRY FESTIVAL EDITION

THE PLANT CITY COURIER

Published Each Tuesday and Friday in Plant City, the Largest Inland Shipping Point in Florida

PLANT CITY, FLORIDA, FRIDAY, MARCH 1, 1940

Catherine Fletcher Named 1940 Festival Qu

Queen Catherine Fletcher

Weather To

Baby Parade

Florida Strawberry Festival Program

City Takes

Winners In

A special March 1, 1940, "Florida Strawberry Festival Edition" of the *Courier* covers the festival's events.

1941

Moving to the new fairgrounds was another step in the growth of the festival, and it was proving auspicious. Despite distant rumblings of war in Europe, going into 1941 things were looking very good for the festival association. The directors and general manager were experienced and planning went well. For the second time, the Grand Parade formed on East Reynolds, marched through downtown, and headed west to the new fairgrounds.

At 2:30 p.m. on Tuesday, March 4, the marchers and spectators could attend the formal opening of the festival, which included Master of Ceremonies Chamber President Arthur Boring, Festival Association President George Carey, Plant City Mayor Don Walden, and the special honored guest, Tampa Mayor Robert E. Lee Chancey. At sixty years of age, Chancey, a proud member of the Knights of Pythias, was serving his tenth year in office. Although Tampa was fraught with corruption, Chancey was acknowledged as an outstanding attorney. There was no admission fee this night, and the Silver Blossom Shows, special acts, and midway attractions went on until midnight.

Another Tampa official was honored on Wednesday, March 5, which was announced as Brorein Day in honor of Carl D. Brorein, president of the Florida State Chamber of Commerce, president of the Florida State Fair Association, and president of the Peninsular Telephone Company, which was the company servicing Plant City. Brorein was the main speaker at the noon luncheon at the Hotel Plant. His uncle, William G. Brorein, founder of the Peninsular Telephone Company, had been the special guest at the 1933 festival.

The comedy parade of Dokkies opened the events on Thursday, March 6, ending in time for the "Impromptu Program" at the grandstand at 2:30 p.m. From 7:30 to 8:00 p.m., prior to the grand coronation, a special entertainment program featured Frankie Connors, "the Silvertone Tenor of Radio."

Queen Jane Langford and her court are ready to ride in the 1941 Grand Parade.

Thursday evening, seventeen-year-old lovely brunette Miss Jane Langford was crowned by Mayor Walden in an elaborate ceremony that included six maids, two heralds, a crown bearer, two flower girls, and three train bearers. The crowd was recorded as the largest coronation crowd in the history of the festival. Both grandstands were filled and the grounds were crowded with people who stood throughout the ceremony. And the weather was quite favorable.

There was a ten-cent fee for park admission and an additional twenty-five-cent admission for the coronation that night. It is interesting that the balloting for the queen and maids, as well as the coronation ceremonies, were under the sponsorship of the newly formed Plant City Quarterback Club, which was raising funds to help finance a gymnasium and civic center for Plant City.

It is also notable that the queen contest rules stipulated to be elected queen or maid, a contestant must receive a minimum of one thousand votes. (The process still counted votes as a penny a vote, and a voter had to submit the penny along with the written name of the candidate and deposit it in one of the many ballot boxes stationed at various locations in Plant City.) In 1941, after the queen was elected, only six other candidates received the requisite one thousand votes. The contest ran from

Saturday, February 15 to Thursday, February 27, when at 6:00 p.m. the new queen and her court would be announced by the Queen's Contest Committee meeting at the Hillsboro State Bank. At 9:00 p.m. the 1941 queen-elect and the six court members, rather than the usual eight, were publicly introduced to a capacity crowd at a ceremony onstage at the Capitol Theatre and were received with excited applause.

Friday, March 7, was an event-filled day that included both the highly popular Baby Parade and the Governor-Cabinet Day—special this year because Governor Spessard L. Holland was from Bartow and was well liked and well connected in Plant City.

The Baby Parade, with sixty-eight excited participants, began at 3:15 p.m. and the youthful marchers paraded through Adelson Field to the grandstand, where the judges would present ten loving cups and thirty ribbons to the jubilant youngsters and their parents. Admission was twenty-five cents plus an additional ten cents for the grandstand itself.

The evening event was a special six o'clock banquet at the Hotel Plant, sponsored by the Kiwanis Club, the Lions Club, and the Florida Strawberry Festival Association in honor of Governor Holland and his cabinet members, many of whom were in attendance. Holland was introduced by Arthur Boring, a close friend and a colonel on the governor's staff. Holland, a decorated veteran of World War I, praised Plant City's productive farmers and spoke briefly on national defense. Over 170 attendees filled the dining hall.

Closing day, Saturday, March 8, included the 2:30 p.m. "Impromptu" ceremonies at the grandstand and a combination of acts at 7:30. Admission for the evening was ten cents. Wrapping up was the big event was the American Legion car raffle, with no warning of the years of war to come or the six-year hiatus for this grand Plant City celebration.

What happened next was a surprise. The State of Florida was planning a Florida exhibit at the Miss America pageant in September in Atlantic City. Jane Langford, Florida Strawberry Festival Queen, was invited to participate in the events in Atlantic City and to ride on the Florida float. Along with Miss Florida, the Orange Festival Queen, and the Honey Festival Queen, Queen Jane Langford, in a striking white bathing suit, rode atop the award-winning Florida float on Tuesday, September 2, 1941, in the ten-mile Miss America Beauty Pageant Parade in Atlantic City before 50,000 cheering spectators. It was a day she has never forgotten.

But it wasn't over yet. With the attack on Pearl Harbor on December 7, 1941, the United States entered the Second World War. Men were being drafted by the thousands and many more rushed to sign up. Events were cancelled across the board, and the Florida Strawberry Festival was put on hold. Even so, in March of 1942, with no festival and no Grand Parade, the Lions Club sponsored "the first big rodeo ever held here" and organized a "gigantic street parade" on Friday, March 6,

Jane Langford, the 1941 queen, sits astride her favorite steed in the Lions Club rodeo in 1942.

that ended at the rodeo grounds at Adelson Field. Mayor Chancey and Mayor Walden rode side by side; Queen Jane rode her horse beside her father, L. T. Langford; and the maids followed in a horse-drawn white wagon.

Jane Langford married Harrison Wall Covington and continued to represent the Florida Strawberry Festival for seven years, until the American Legion restarted the festival in March of 1948, after the conclusion of the war.

1948

Plant City suffered heavily as most small communities did in the war years. Men were drafted virtually off the streets, and towns were quickly becoming like ghost towns. When they returned, the Plant City area men joined the American Legion, which was formed after World War I. The local post was named for Norman McLeod, the first Plant City fatality of that war. After World War II, the Norman McLeod Post 26 rapidly grew to over five-hundred members, in a community with a population of approximately 8,800.

There was talk about reviving the strawberry festival but the Florida Strawberry Festival Association had not yet reorganized. In December 1947, according to the foreword in the 1949 program (signed by the fifteen members of the board of directors), "The Strawberry Festival Association decided to turn over the operation and renewal of the event to the Norman McLeod Post No. 26, of the American Legion. With a membership in excess of 500, the Legion post constituted the largest civic organization in the city."

Festival Association Managing Director Gerald H. Bates continued, saying that while the American Legion is sponsoring the festival,

> the annual Festival is a community-wide event and every effort has been made to enlist the aid, assistance and co-operation of the community as a whole in staging this annual exposition . . . We, the Directors of the Plant City Strawberry Festival Association, all of whom are members of Norman McLeod Post No. 26, of the American Legion, are endeavoring to do a good job. We are presenting what we feel to be a truly representative picture of Plant City and East Hillsborough county activities, and it is our sincere wish and hope that this will prove enjoyable to you.

A *Courier* article on Thursday, February 26, announced, "STAGE NOW SET FOR PLANT CITY BERRY FESTIVAL." The 1948 revival of the Florida Strawberry Festival, after being dormant since March of 1941, would run for five days—Tuesday, March 9 through Saturday, March 13. Restarting the festival was facilitated by the location of the American Legion property, the access to the armory complex, and the adjacent

The Florida Strawberry Festival was reinvigorated in 1948 by the Norman McLeod Post 26 of the American Legion in Plant City.

city-owned Adelson Field. The festival had relocated there in 1940, making it a permanent site for the event.

Restarting the festival meant a lot of planning had to go into the process and decisions made as to all the details—how many days should the festival run, format for the queen contest, speakers, the parade, exhibits, Baby Parade—every detail of the previous festivals had to be revisited and decisions had to be made for this, the thirteenth strawberry festival. It was to be "the most elaborate festival yet held."

The first big change came with the format for the queen contest. It would be judged by a panel of judges and no longer would be open to balloting by the public at a penny a vote. This year the contest was set for Friday night, February 27, at the Plant City High School auditorium. The queen and maids would be selected from the thirty-one contestants entered in the contest; the coronation was to take place Thursday, March 11, at Adelson Field. Another change was, for the first time, a cattle exhibit would be held with the festival.

The Grand Parade, scheduled for Tuesday, March 9, was to have a complete lineup, according to the news article, "including 11 out-of-town floats, four bands, a drum corps, and a number of decorated cars by local business firms." The parade route also was changed to form on Haines Street (now Dr. Martin Luther King, Jr., Boulevard), proceed east to Collins Street, north to Reynolds Street, and straight west to the new festival grounds.

Little is known about some of the days' activities for the 1948 festival. There was a Grand Parade, some exhibits, and a queen contest and coronation, but little information exists about the speakers and other special events. The coronation ceremony, however, was spectacular.

The thirteenth annual queen coronation was set for Thursday, March 11, at 8:00 p.m. on the newly constructed stage. Like a stage play, the program had three scenes: Scene I was the entrance of all the royalty, the band, and the Colors. The singing of the "Star Spangled Banner" followed the coronation ceremony. "Happy over the return of their Strawberry Queen, the children of East Hillsborough County celebrate" with tumbling, dancing, and choral selections (making up Scene II). During Scene III, the entire audience joined in "God Bless America" while the ensemble in recessional left the stage.

The first post-World War II Florida Strawberry Festival saw Barbara Alley crowned queen in 1948.

The 1948 festival queen, court, and entire cast attend the coronation ceremony at the new fairgrounds on West Reynolds Street.

The new festival queen, with her escort, ascended the throne, attended by her four maids and their escorts. They were followed by "outstanding boys and girls selected by their fellow students at the Plant City Schools." Next came the Plant City High School Band, the Colors, and the flower girls, crown bearer, and train bearers, who were from Bryan, Burney, Wilson, and Jackson Elementary Schools.

At this point, Miss Barbara Alley, the lovely daughter of Mr. and Mrs. Fred Alley, received the crown relinquished by the 1941 queen, Mrs. Harrison Wall Covington (Jane Langford), from Mayor Lennox E. Morgan. Joining in this ceremony were the junior maids and escorts from Plant City Senior High School and the ladies and lords-in-waiting from the elementary schools. In all, there were seventy-eight persons participating. It was a grand and glorious night.

The 1948 festival probably closed Saturday night, March 13, after the American Legion car raffle. It was a grand and successful new beginning of the tradition of the Florida Strawberry Festival in Plant City.

1949

From the beginning of the effort to revive the festival, it was clear the American Legion had the enthusiasm and the manpower to get things done. The front of the souvenir program referred to it as the Plant City Strawberry Festival, sponsored by the Norman McLeod Post No. 26, American Legion. The association letterhead read: "Florida Strawberry Festival, Inc., at Plant City, Fla.," and below it was the reference to being sponsored by the local American Legion Post 26.

In the 1949 program was also a statement that the Plant City American Legion Post 26, sponsor of both the 1948 and 1949 festivals, had pledged $16,000 to the Baptist Hospital Fund and 25 percent of the profits from the 1949 festival activities would be given toward the hospital building fund, thus "helping the much-needed hospital to early completion."

February 24, with the March 1 festival opening approaching, Festival Manager Gerald Bates wrote Dr. Doak Campbell, Florida State University president, on behalf of three FSU students: 1948 Queen Barbara Alley, 1948 maid Eldora Holsberry, and 1949 maid Peggy Sparkman. Bates requested they be given a waiver to take required tests before or after the festival. He wrote, "They are earnest students and want to get back as soon as possible and, . . . we will do everything in our power to expedite their return to school." It worked.

The postwar years had been good for the economy and the fourteenth annual Florida

Strawberry Festival was predicted to be one of the best. This year a special luncheon for distinguished guests was held at 12:30 p.m., before the Grand Parade, which began at 1:30. The parade route changed again and, after forming east of the SAL tracks on Reynolds Street, paraded through several city streets before heading west to the fairgrounds.

At three o'clock the formal opening of the festival featured Plant City Mayor Henry S. Moody, along with Tampa Mayor Curtis Hixon, several county commissioners, and other officials. The festival officers also presented the awards for the winners in the parade. Another change this year was that at 4:00 p.m., contests were held for junior drum majors and for baton twirlers, the winners of which would be awarded cash prizes.

As usual, free acts went on at various times of the day and the Royal Crown Shows presented its midway attractions and shows, lasting until midnight. Additionally, the festival association had booked the Renfro Valley show, a nationally popular radio country music performance. It was advertised as "one of our main performances and will be shown at remarkably cheap admission." The show was booked every day except Coronation Day.

American Legion Day was Wednesday, March 2, and honored all veterans who served in the world wars. The luncheon was at the American Legion Home, adjacent to the fairgrounds, and featured distinguished guests and Arthur Boring, past Legion state commander, as emcee. Speeches were presented in the afternoon.

Governor's Day and Coronation Day followed on Thursday, and a banquet honoring Governor Fuller Warren was set up by J. Arden Mays, the governor's "No. 1 local colonel," for 6:30 p.m. at the Hotel Plant. Warren had served in World War II as a gunnery officer and had just taken office as governor in January.

The coronation ceremony was as elaborate as ever. The Lions Club and Auxiliary had run the selection process and, at the pageant in the Plant City High School auditorium on February 11, Miss Peggy Hodges, petite daughter of Mr. and Mrs. William C. Hodges and a junior at Florida

Queen Peggy Hodges and train bearers promenade to the stage at the fairgrounds in 1949.

Barbara Alley, 1948 festival queen, congratulates the 1949 queen, Peggy Hodges.

These children are all dressed up for the 1949 Baby Parade.

Southern College, was chosen out of seventy-three contestants. With all the pageantry and trappings of royalty—six maids and a covey of heralds, pages, train bearers, and a crown bearer—Hodges was crowned festival queen by Governor Fuller Warren. Warren arrived late for the dinner, having flown into Drane Field from Miami then motoring to Plant City, but he was on hand to crown the 1949 festival queen. Miss Barbara Alley, 1948 festival queen and a student at Florida State University, greeted Hodges with a smile and a hug.

The Kiwanis and Lions Clubs had planned a joint luncheon for Friday at 12:30 p.m. for all local civic clubs at the Hotel Plant in honor of Governor Warren, who was also to be the principal speaker. Unfortunately, Warren did not make it to the luncheon, having already flown from Drane Field back to Tallahassee. The Friday, March 11, *Courier* headlines read: "The Gov. Wasn't There, But 148 Ate Turkey At Luncheon In His Honor."

The feature event on Friday was the always popular Baby Parade, with a line of youngsters, toddlers, and floats streaming across the stage at Adelson Field. The usual collection of loving cups and ribbons were presented to the many excited children and parents. Saturday, the last day of the festival and the last festival of the decade, there was a luncheon for the Honorable Nathan Mayo, the well-liked commissioner of agriculture, who brought several officials from the US Department of Agriculture as his guests. And the closing event was the American Legion car raffle where a brand new Dodge was awarded to a lucky ticket holder.

After the festival ended, Queen Peggy Hodges had several exciting events. One was her marriage to William Roauer in June 1949. Then in December, compliments of Eastern Air Lines, Queen Peggy Hodges Roauer flew to New York City to deliver the "first pint" of Plant City strawberries to Mayor William O'Dwyer, a general during World War II and the one-hundredth mayor of New York City. O'Dwyer was in the hospital at the time and she delivered the strawberries to Acting Mayor Vincent Impelliteri, compliments of Plant City Mayor Henry S. Moody. It was a thrilling year for the new festival queen and the end of a truncated decade for the Florida Strawberry Festival.

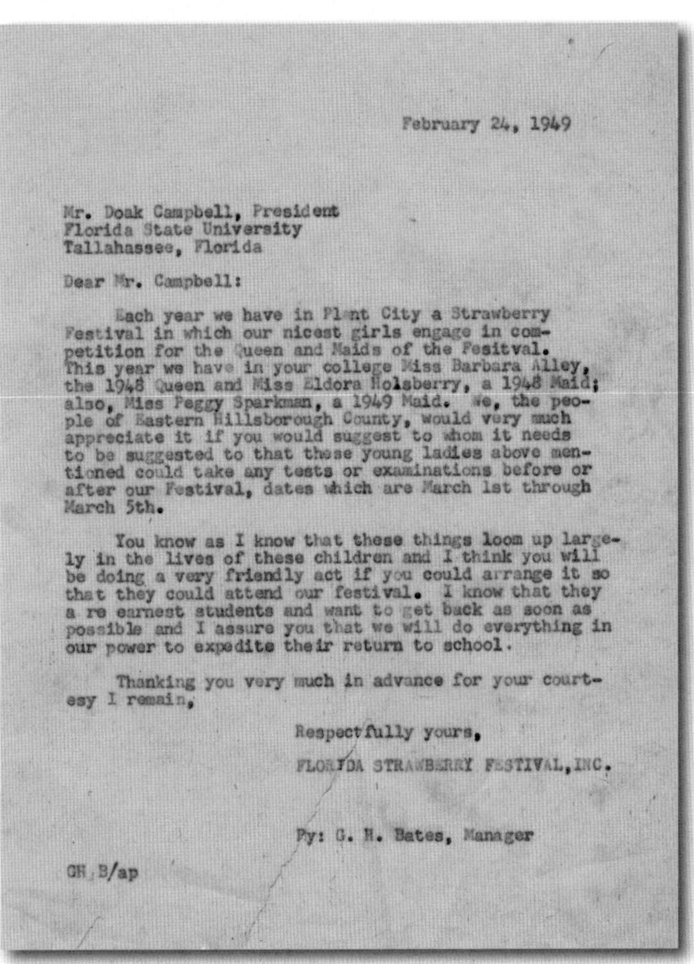

February 24, 1949

Mr. Doak Campbell, President
Florida State University
Tallahassee, Florida

Dear Mr. Campbell:

Each year we have in Plant City a Strawberry Festival in which our nicest girls engage in competition for the Queen and Maids of the Fesitval. This year we have in your college Miss Barbara Alley, the 1948 Queen and Miss Eldora Holsberry, a 1948 Maid; also, Miss Peggy Sparkman, a 1949 Maid. We, the people of Eastern Hillsborough County, would very much appreciate it if you would suggest to whom it needs to be suggested to that these young ladies above mentioned could take any tests or examinations before or after our Festival, dates which are March 1st through March 5th.

You know as I know that these things loom up largely in the lives of these children and I think you will be doing a very friendly act if you could arrange it so that they could attend our festival. I know that they a re earnest students and want to get back as soon as possible and I assure you that we will do everything in our power to expedite their return to school.

Thanking you very much in advance for your courtesy I remain,

Respectfully yours,

FLORIDA STRAWBERRY FESTIVAL, INC.

By: G. H. Bates, Manager

GHB/ap

Florida Strawberry Festival letter to FSU president.

THE 1950S: A DECADE OF POSTWAR NORMALCY AND GROWTH

The decade of the 1950s saw many changes amidst economic growth and a conservative social environment. The Truman years ended in January of 1953, followed by the July 27 signing of the armistice that brought the horrifying Korean Conflict to an end. The Cold War set in, the nation's hero and avuncular leader Dwight D. Eisenhower was elected president, and *Sputnik* brought a breakthrough and renewed interest in technology.

In the 1950s the overall festival acreage continued to grow. The festival took over the National Guard Armory complex, swallowed Adelson Field and the former American Legion property, and bought more land for parking. It expanded its exhibits and added more vendors and food purveyors, but the festival never included alcohol. They also continued the tradition of themed days and extended the festival to six days, running Monday through Saturday—never on Sunday.

Coming out of the war years, Gerald H. Bates, the association chairman and long-time leader, became the festival general manager for the 1948 and 1949 festivals. By 1950 that role was assumed for a time by G. R. Patten. The chairman of the festival association was Elmer N. Dickinson, who was the manager of the W. W. Mac Department Store downtown and later became the area supervisor for the State Revenue Commission. Dickinson was the festival association chairman/president from 1950 until 1972. By the mid-1950s Fred W. Nulter, a builder and contractor, began filling the role of general manager and by 1957 he officially held that title.

The period from 1950 to 1960 also saw the second-greatest population growth in Plant City. The city swelled from 9,230 in 1950 to 15,711 by 1960, a 70.2 percent increase. The economy was doing well and the city and surrounding communities prospered. The American Legion remained strong and was the leading force for the festival through the decade, with every member of the festival association board of directors also being members of the Norman McLeod Post No. 26, American Legion.

1950

The 1950 festival (February 27 through March 4) began several changes in the festival schedule. There had been some discussion about the queen and court contest, or pageant, being held a week or so prior to the festival and in a different venue. They added Monday to the schedule for the queen pageant itself, taking place on the outdoor stage at the festival grounds. In 1950 forty hopefuls crossed the platform that chilly evening in front of many excited spectators and three judges. Etta Mae Helzer, of Plant City, emerged as the queen-elect for the 1950 festival. The queen contest and coronation were organized by the American Legion Auxiliary.

The Grand Parade was set for Tuesday, February 28, and this was the first time the big luncheon was held prior to the parade. It has become a tradition today, although moved to an earlier time to allow the attendees to get to the parade on time. In 1950, at noon, nearly eighty county and municipal officials, civic leaders, and festival officials were guests at the luncheon at the Hotel Plant. They then headed for the Grand Parade.

The 1950 parade was called the "Best Ever" with a record crowd gathered on the city streets to enjoy the grand event. The *Courier* reported, "The sun-spangled day was made to order for the holiday crowd who jostled one another in the carnival spirit."

Queen Etta Mae Helzer and her court ride in the 1950 Grand Parade.

After the parade, there was another luncheon, this one being held at the adjacent American Legion Home in honor of local and visiting American Legion officials. As it was for the remainder of the decade, Tuesday was also American Legion Day and Children's Day, wherein Legionnaires wearing their caps and children up through high school were admitted free.

Wednesday was traditionally Agriculture Day and the 1950 festival followed suit. Agriculture Commissioner Nathan Mayo was the honored guest.

Thursday was Coronation Day. The 1950 ceremony was highly regal and courtly with Queen-elect Etta Mae Helzer and her court of eight maids, eight escorts, two heralds, two train bearers, one ring bearer, four flower girls, and additional courtiers. In all, twenty-nine participated in the ceremony. They were joined by dancers, the Plant City High School Band, and several festival queens from years past. Mayor Henry Moody was proud to place the shining and splendid crown on Queen Etta Mae's head. It was an extraordinary evening.

The Friday Baby Parade was moved this year to circling the City Hall Park on Mahoney Street, starting on Evers Street, south to Reynolds, east to Collins, and ending up at the reviewing stand in the park for awards. The last day of the festival, Saturday, was named All Hillsborough Day and Tourists Day. There was a luncheon for visiting dignitaries. Once again, the finale was the American Legion automobile raffle, which concluded the annual festival.

Three former Florida Strawberry Festival queens are honored onstage at the 1950 queen coronation ceremony.

1951

In 1951, after using Royal Crown Shows for several years, the festival contracted with Prells Broadway Shows for the shows and midway. The association later came back to Royal Crown, then contracted with Blue Grass Shows for the remainder of the decade. The midway rides and shows were still carnival-like, but the association made it clear this was to be the same family-oriented environment still prevalent today.

The festival (March 5–10) again began on Monday, with the main event being the queen and court selection process. Thirty-nine contestants were interviewed and reviewed as they strolled by in evening gowns and in swimsuits on an outdoor stage on March 5! It was cold, but the décor was elaborate and there was entertainment for the crowd of 1,200 who watched silently as the judges carefully selected the new queen. The lovely blonde Plant City High School student Edwana Snowden, and her beautiful court, were crowned at the ornate Thursday night ceremony. This year the Elks Lodge oversaw the selection process and the coronation.

Tuesday, March 6, was American Legion Day and Children's Day—and the Grand Parade. The "informal get together" for prominent guests and officials at the Hotel Plant was moved up to 11 a.m. and the parade started at 1 p.m. Legionnaires and children through high school were admitted free, but the rides for children were nine cents a ride. The 1951 queen-elect and her court and the 1950 reigning queen and her court all rode in the parade through downtown and west to the festival grounds. The *Courier* carried it this way: "A capacity crowd of uncounted thousands of home folks and tourists jammed Plant City's streets Tuesday afternoon to watch the colorful 16th annual Strawberry Festival parade. Seventeen floats were decked with pretty girls in frothy gowns and bathing suits, nine bands boasted bright uniforms, strutting majorettes and baton twirlers from elementary schools added color in their costumes . . . in the hour-long parade."

Wednesday the Kiwanis and Lions Clubs joined in hosting a luncheon at the Hotel Plant in honor of Agriculture Commissioner Nathan Mayo, one of Plant City's famous friends in Tallahassee, and included the reigning queen and court, the queen-elect and her court, civic leaders, and festival officials. The colorful coronation was the featured event for Thursday night and included fanfare, elaborate decorations, music by the high school band, and the many courtiers. The proud Mayor Lee Slaght placed the crown on Miss Snowden to the applause of the hundreds of spectators. This year the price of admission to the coronation ceremony was twenty-five cents.

Friday, March 9, was Tourist Day and this year there were prizes for oldest, farthest traveled, and several other categories. The next day was the final day and concluded with the usual American Legion car raffle, the proceeds from which would be donated to a local charitable cause.

Queen escorts Cecil Everidge and John Hutchinson proceed to the stage for the coronation ceremony.

The newly crowned 1951 queen, Edwana Snowden, is shown with her court at the coronation.

1952

The seventeenth annual Florida Strawberry Festival was held from February 25 through March 1 of 1952. It was a leap year and Friday, Baby Parade Day, fell on February 29. Monday the 25th was the quiet opening of the festival and things were slow until the lights came on that night for the queen pageant. The parade of young girls passed by the judges and by the end of the night a new queen and her six maids had been chosen. (The details are incomplete. The queen selection process changed in 1952 or 1953. It was about this time the Junior Chamber of Commerce took over the queen selection and coronation proceedings.)

The Grand Parade stepped off Tuesday at about 1:00 p.m. after the informal get-together sponsored by George Carey at the Hotel Plant. The festival's own multi-level, flower-festooned float carried the 1951 queen, Edwana Snowden, and her eight maids. It was a beautiful cool but sunny day as the Plant City High School Band paraded through the downtown streets led by Drum Major Donald

Downtown Plant City is on the parade route for the 1952 Grand Parade.

The Grand Parade is a spectacular show as it passes through downtown on its way to the fairgrounds in 1952.

Rayburn with eight majorettes close behind. They marched west, all the way out Reynolds Street, and passed the American Legion Home then dispersed. It was another highly successful parade. It was also American Legion and Children's Day and Legionnaires wearing their caps and children through high school were admitted free to the festival grounds.

Wednesday was Agriculture Day and State Agriculture Commissioner Nathan Mayo would be the guest of honor. Then, on Thursday, the main event was the ceremonial coronation of Queen Arvor Lois Harris, the glamourous seventeen-year-old daughter of Fred and Lonnie Harris of Lacoochee in Pasco County, Florida, now living in Brandon. Her court consisted of six maids: Ann Brown, Nell Jernigan, Elizabeth Sloan, Dottie Weldon, Barbara Parish, and Sandra Gentry.

The annual Baby Parade stepped off on Friday, February 29, and the children again paraded through the downtown streets, some on foot and most on elaborate miniaturized floats. They headed to the reviewing stands and received their awards in City Hall Park on Mahoney Street. The final day of the festival for 1952 was Saturday, March 1, and the main feature, after the luncheon that day, was the American Legion raffle that night. Then the shows, rides, and midway closed the festival at midnight.

1953

The eighteenth annual Florida Strawberry Festival (February 23–28, 1953) opened with few changes. The Royal Crown Shows again provided "thrilling rides, fifteen new exotic shows, and sensational circus free acts," all on the midway!

The queen selection and coronation were now under the direction of the Plant City Jaycees, with Hilman Bowden, Cecil Everidge, and Lester "Buddy" Blain in charge. The first change they made was to select, from the many contestants, the seven finalists for the queen pageant on Monday night and then select the queen on Thursday night at the coronation ceremony itself. Of the other six finalists, now maids, one was chosen as first maid. So it was that on Monday, February 23, the girls vying for the festival queen's crown strolled across the platform at the festival grounds under the lights as the three judges reviewed their attributes. The seven finalists chosen were delighted, then turned to the necessity of preparing for the Grand Parade the next day.

The Grand Parade was set for 1:00, following the 10:30 a.m. informal get-together of special guests sponsored by George Carey at the Hotel Plant. For some reason the parade route reverted to the one used years ago and formed on East Reynolds Street east of the SAL tracks and proceeded to wind its way through the downtown streets before heading west on Reynolds to the festival grounds.

An aerial provides a bird's-eye view of the Florida Strawberry Festival grounds in 1953.

As usual, the parade route was lined with hundreds of excited spectators stretching to see all the floats, marching bands, dignitaries' cars, and other marching units. At the festival grounds Legionnaires wearing their caps and children got in free, as well as all the parade participants.

The only special event on Wednesday, February 25, Agricultural Day, was the Kiwanis-Lions luncheon in honor of Florida State Agriculture Commissioner Nathan Mayo. Additional attendees included state cabinet members, city commissioners, the reigning queen and her court, queen candidate finalists, association directors, and other guests.

The selection and coronation of the 1953 festival queen was held on Thursday night, February 26, under the lights. It was another royal show, with the reigning queen and court, the seven anxious candidates, heralds, train bearers, courtiers, and flower girls, all amidst flowery and glittering decorations. From amongst the seven emerged vivacious seventeen-year-old Plant City High School senior Ruby Jean Barker, daughter of Mr. and Mrs. E. C. Barker.

Miss Barker was crowned by Mayor L. E. Morgan Jr. "during a brilliant coronation pageant on the Festival grounds here last night," said the *Tampa Daily Times* (Friday, February 27, 1953). "The blonde teen-age queen was selected from a field of seven finalists before a crowd of some 1500 persons who filled the grandstands to witness the event."

The Friday Baby Parade started at 3:30 p.m. and the babies, toddlers, and young children strutted around the block in their costumes and outfits and some atop their floats. After passing the review stand in City Hall Park, the smiling youngsters and their parents received their awards.

"All Hillsborough and Tourist Day" was the theme of the final festival day on Saturday, February 28, and the special event of the day was the 3:30 p.m. Tourist Get Together with free acts and tourist prizes. The night closed out with the American Legion car raffle and the lights went off at midnight when the Royal Crown Shows and midway closed down.

Mayor L. E. "Rat" Morgan is all smiles as he places the crown on Ruby Jean Barker, 1953 festival queen.

1954

The 1954 festival, still sponsored by the Norman McLeod Post No. 26, American Legion, ran from Monday, February 22, through Saturday, February 27. Opening on Monday, the main event was the selection of the seven finalists for the queen pageant, which took place that night on the outdoor platform at the festival grounds. It was directed by Buddy Blain and others of the Plant City Jaycees.

Tuesday was the Grand Parade, with the usual George Carey-sponsored get-together at the Hotel Plant preceding the parade. The day was blessed with ample sunshine and massive beautiful floats, such as those of the Seaboard Air Line Railroad and Sears Roebuck and Company, which streamed through the city streets while throngs of wide-eyed spectators enjoyed the bands and marching units.

Wednesday, Agriculture Day, again paid tribute to Plant City's sustaining crop—King Strawberry! Fittingly, the guest of honor was State Agriculture Commissioner Nathan Mayo and an assortment of dignitaries, queen pageant participants, civic leaders, and festival officials.

Queen selection and coronation were held on Thursday, February 25, and the Jaycees put on a spectacular presentation on the festival stage, with the usual regally attired heralds, train bearers, flower girls, and many courtiers, amid the flowers and glitter. The Plant City High School Band played and the judges reviewed the final seven, noting their poise onstage and in the interviews. The judges chose Ruth Shuman of Springhead, the eighteen-year-old daughter of Mr. and Mrs. Foy A. Shuman and a senior and queen of Plant City High School, as the queen of the nineteenth annual Florida Strawberry Festival. Plant City

The beautifully detailed Seaboard Air Line Railroad float looks great as it passes through downtown Plant City in the 1954 Grand Parade.

Mayor Otis M. Andrews, long-time general manager at McCrory's Five and Dime, proudly placed the crown on an excited and nervous Queen Ruth. The band played in celebration.

The Baby Parade on Friday followed the downtown route on Mahoney Street and passed the review stand at City Hall Park, with the winning children and parents receiving their colorful loving cups and ribbons on the stage. On Saturday, February 27, the festival closed with a get-together with acts and prizes for tourists and the grand raffle of the new automobile that night.

Plant City Mayor Otis Andrews proudly crowns Ruth Shuman as the 1954 Florida Strawberry Festival Queen.

1955

The *Tampa Tribune* reported the twentieth annual Florida Strawberry Festival (February 21–26, 1955) in this way: "Every phase of agriculture in Hillsborough County went on display today as the 20th annual Florida Strawberry Festival opened in Plant City, but the spotlight was on the strawberry, this area's major crop for 50 years. Every community in the area was represented in the elaborate exhibits, many of them presented by rural youth organizations."

Suspense rose as Monday evening's selection of queen finalists began at 7:30. James Quarles was the Jaycees' master of ceremonies. The judges measured the thirty-nine swimsuit-clad candidates who strolled across the platform in the cool night air and, after a period of entertainment and consultation, selected seven finalists for the Thursday night final queen selection and coronation ceremony. Rosemary Hardy, Margaret Noe, Mary Jane Kennedy, Bobby Jean Harwell, Shirley Brown, Betty Clements, and Phyllis Hobkirk were guaranteed at least a maid selection. Admission to the night's event was now thirty-five cents.

The Grand Parade on Tuesday followed the earlier get-together of dignitaries and special guests at the Hotel Plant sponsored by former association president George Carey. Stepping off at 1:30 p.m. on another sunny day, the parade followed the old route lining up east of the railroad tracks on Reynolds Street, winding through downtown streets, and ending up this year at the new William Schneider Memorial Stadium, adjacent to the armory complex and the American Legion Home. Thousands of fans, local residents, and tourists alike, cheered as the many floats, bands, marching units, and decorated cars passed by.

Agricultural Day was Wednesday, February 23, and the Kiwanis-Lions luncheon in honor of Agriculture Commissioner Nathan Mayo grew to include numerous officials and festival participants. The toastmaster was Mayor Otis M. Andrews.

The sponsor of this year's festival was the American Legion. Fred Nulter was the secretary-manager of the festival, Buddy Blain of the Junior

A marching band in the 1955 parade enters the Florida Strawberry Festival grounds and the new Schneider Memorial Stadium.

Betty Clements, 1955 festival queen, and her court at the ornate coronation ceremony.

Chamber of Commerce was the chairman of the queen selection and coronation, and the Blue Grass Shows ran the midway.

At 8:00 p.m. Thursday, February 24, the selection and coronation ceremony began at the festival stage in front of hundreds of enthusiastic onlookers, fans, friends, family, and tourists. John Taylor, of the Jaycees, was the emcee and admission was thirty-five cents. The elaborate program began with the procession of the 1954 festival queen and her court, followed by the presentation of the judges to her majesty Queen Ruth by city attorney Paul Buchman. Next came the presentation of the candidates for selection at which time the seven finalists paraded before the three judges and carefully responded to the judges' interview questions.

The audience was treated to entertainment as the judges consulted. Queen Ruth then gave her farewell proclamation and abdicated her throne as the emcee announced the new queen, who was then crowned by Mayor Andrews. After excited congratulatory applause came the procession of the new festival queen, Betty Clements, and her first maid and court, who were presented to the vast audience. The photo of 1954 Queen Ruth Shuman placing the tiara on Queen Betty Clements was wonderful and hit the news wires, being picked up across the country.

The Baby Parade this year was changed from downtown to staging at the new William Schneider Stadium adjacent to the festival grounds, and the gaily dressed children passed by the review stand and onto the platform for their prizes. Saturday, February 26, concluded the festival and featured the tourist get-together, acts, and prizes. The American Legion new automobile raffle was held late that night, and the Blue Grass Shows closed the twentieth annual festival at midnight.

In a photo that went nationwide, Ruth Shuman, 1954 queen, passes the crown to Queen Betty Clements in 1955.

1956

The 1956 festival (February 20–25) got off to a quick start with thirty-eight girls contesting for the honor of becoming the Florida Strawberry Festival Queen for the following year. The Plant City Jaycees were in charge of the twenty-first annual queen selection and on a cold Monday night, February 20, Virginia Young was selected as queen-elect, along with first maid-elect Mary Jane Jackson and maids-to-be June Fussell, Delores Harris, Carolyn Hardee, Maureen Crum, and Marie Stanphil. They would all ride in the Grand Parade Tuesday afternoon and looked forward to the glamorous coronation on Thursday night.

The Grand Parade on Tuesday, February 21, marched through downtown streets to the delight of the hundreds of spectators. Among the floats, cars, bands, and marching units were the young girls of Brownie Troop 243 from Jackson School and the Springhead Girl Scout Troop 21, both groups riding on locally owned commercial trucks. Representatives of the Plant City Youth Association passed by in a 1956 Ford convertible. Governor LeRoy Collins rode in a new Oldsmobile convertible driven by Ben Rawlins.

Coronation evening, Thursday, February 23, soon followed and the Jaycees-organized ceremony included musical entertainment, dancing, glitter, and royal decorations. Courtiers tended to the newly crowned festival queen, Virginia Young, and her court.

The biggest event of the 1956 festival was the popular Baby Parade held on Friday, February 24. This year drew 130 entrants, all of whom were in homemade intricate costumes and most riding miniature floats, as they passed the review stand and a large crowd of applauding spectators in Schneider Stadium. This year the queen and king and their court were crowd favorites, and the photo of Queen Janice Hutchinson and King Randy Tickel in a celebratory kiss made the rounds and the newspapers.

The lights went off after the American Legion raffle on Saturday night, February 25, and the 1956 festival came to a close. But more was to follow.

A bevy of beauties takes the cool outdoor stage in the 1956 Florida Strawberry Festival Queen Contest preliminary.

In 1930 the criteria for entering the queen contest were liberal about age and marriage. Contestants included misses and matrons of various ages. In 1949, Queen Peggy Hodges became Mrs. William Roauer in June of 1949 and completed her year of service. In 1956 that provision had apparently changed, and when Queen Virginia Young married Donald Beery that June she relinquished her crown and the festival association asked first maid Mary Jane Jackson to complete Queen Virginia's term.

The Baby Parade queen and king share a congratulatory kiss in 1956.

Later that summer, officials of the National Strawberry Queen Contest of Glenwood Springs, Colorado, invited the Florida Strawberry Festival to participate in its Strawberry Days celebration and National Strawberry Queen Contest in August. The Jaycees promoted the opportunity and encouraged Mary Jane Jackson to enter the contest. The *Tampa Daily Times* reported the story and added that on July 30, Miss Jackson was the guest of Dick Pope, owner of Cypress Gardens, for a tour and publicity photographs.

With full Jaycees support, L. M. Buddy Blain recruited the necessary funding from the county, city, Plant City Chamber of Commerce, and the Florida Strawberry Festival. Wednesday morning, August 8, Miss Jackson with her mother, Mildred Sims Jackson, as chaperone, were on their way to Colorado. After several days of tours, meet and greets, photographs, boat rides, rafting, parades, a rodeo, fish fry, banquets, and dances, the final judging took place on Saturday, August 11, at the historic and elegant Hotel Colorado. Mary Jane Jackson was crowned National Strawberry Queen as the clock approached midnight.

Excited, she immediately called Blain and the Jaycees back in Plant City. It was nearly 2:00 a.m. when she gave them the news. After a week-long all-expense trip to Hollywood, Queen Mary Jane and her mother arrived back in Plant City, where the city council had proclaimed it "Mary Jane Jackson Day."

Glenwood Post 10c

Volume 67—No. 33 | Glenwood Springs, Colo. | Thursday, August 16, 1956 | Ten Pages

Her Majesty National Strawberry QueenMary Jane II

Bill Friend Photo, Plant City, Fla.

Picked at Glenwood Springs was Miss Mary Jane Jackson of Plant City, Florida for National Strawberry Queen for 1956. Mary Jane, 18, was accompanied to Glenwood and on to California by her mother, Mrs. Mildred Jackson.

For a whirlwind tour sponsored by Plymouth Corporation Mary Jane was flown to Los Angeles on a TWA luxury liner. She is staying at the Garden of Allah Hotel on Sunset Boulevard. She reports she has a beautiful room.

She has appeared on Bob Crosby's TV program, Panaramic Pacific Show on KNXT, Johnny Carson's TV show, has been taken on a tour of Universal-International Studios, lunched at the Brown Derby, and visited Disneyland.

She is scheduled to visit Knott's Berry farm, make an appearance at the Hollywood Knickerbocker Hotel, and be a guest at the Pacific Coast Club in Long Beach. Sunday she has free, and will fly back to Denver on Monday.

Florida Entrant Chosen National Strawberry Queen

Lovely Miss Mary Jane Jackson, 18, of Plant City Florida, was chosen by the judges as the National Strawberry Queen for 1956 Saturday night at the big dance.

She is the winner of a trip to Hollywood by plane from Denver as guest of the Plymouth Motor Corporation. She will stay at the Hollywood Roosevelt Hotel and will be entertained while there, going on a whirwind tour of TV and radio stations.

The second place winner was Miss Penny Marrano, 19, of New York and Miss Patricia Lorraine Kelley, 18, of Buchanon, West Virginia was third.

The girls were judged on the basis of 50 points for beauty, 25 for poise, and 25 for personality. Judges Cesare Morganti of Morga[nt]i Modeling Agency in Den[ver]; Lieutenant Governor Steve Nichols; his brother Bill Mc[N]ols; Don Dailey, secretary to [Gov]ernor Ed Johnson; Carleton [...] of the D & R G R. R. in De[nver] and Nello Cassai, Denver Post writer.

Strawberry Queens Parade Saturday In Convertibles

All the beauty queens competing for the 1956 Strawberry Queen title rode along Grand Avenue in convertibles Saturday morning. A big crowd lined the streets.

Also in the parade was Jeanette Podbevsek, a runner-up in the local contest in June.

The girls were followed by their escorts, college boys from the area: Melvin Britton, Miles Anderson, Bob Richmond, Gary Leighton, Jerry Olshove, Doug Jenkinson, Bill Coleman, and Larry Graves.

Plant City Strawberry Festival Queen Mary Jane Jackson was named National Strawberry Queen at a Colorado strawberry festival in 1956.

1957

The growing emphasis on a wider variety of exhibits is noticeable in the 1957 festival, and an example is found in the foreword in the program: "Strawberries constitute a major cash crop in this section, but there are other crops and other industries which contribute mightily to the progress and prosperity of this section. Spring and fall vegetable crops, citrus groves, canning activities, cattle, timber, phosphate—they all play an important role in the lives of those who call Plant City and East Hillsborough their home." The main sponsor of the twenty-second annual Florida Strawberry Festival was still the Norman McLeod Post No. 26, American Legion. E. N. Dickinson remained as president and chairman, and Fred Nulter was the festival's general manager. Dickinson, Nulter, and Cecil Everidge, of the Lions Club, were the central planners for the year's festival.

In the past, there may have been sixteen or more committees; in 1957 these were reorganized into eight: Parade, Building and Grounds, Agricultural Exhibits and Awards, Agriculture, Tourist Events, Public Safety, Entertainment, and Baby Parade, along with the two divisions of the Lions Club and Lions Auxiliary.

The program also carried information about the Junior Agricultural Fair, which was sponsored since 1946 by the East Hillsborough Chamber of Commerce and ran separately from the festival. Although there is more information about the parades, lunches, and queen pageant, there was also continuous activity in the many indoor and outdoor exhibits. In addition, there were many free acts going on throughout the day. Festivals tended to be very busy places.

The 1957 festival ran from Monday, February 18, to Saturday, February 23. It started slowly on Monday, but by 7:00 p.m. it was buzzing in preparation for the opening of the queen selection event where the seven finalists would be chosen. The Lions and Lions Auxiliary organized the selection process, and the review of the forty-three contestants went smoothly, despite the chill of the evening. The judges carefully reviewed each of the contestants' strengths and selected the seven finalists for the Thursday night elaborate and regal queen pageant and coronation.

Tuesday started with the informal get-together at the Hotel Plant, after which the attendees joined the parade or took their spots along the parade route. The parade stepped off at 1:30 p.m. sharp and the marchers strutted from Haines Street to Collins Street, then to Reynolds Street, and headed west to Schneider Stadium and the festival grounds, where they paraded past the grandstands full of excited spectators. As usual, Agriculture Day fell on Wednesday, February 20, and was dedicated to eighty-year-old Florida Agriculture Commissioner Nathan Mayo, as well as "members of his department and members of all agricultural organizations in the state."

The Florida Strawberry Festival Queen Coronation Ceremony, sponsored by the Lions and Lions Auxiliary, would be a spectacular, carefully orchestrated event with former mayor L. E. "Rat" Morgan Jr. as master of ceremonies. The Plant City High School Orchestra played as the 1956 court was presented, followed by a solo number and the presentation of the seven finalists. After more entertainment, the 1956 festival queen and court abdicated the throne and the 1957 court was then introduced, with two heralds, four junior maids, a crown bearer, two senior maid escorts, two flower girls, and two train bearers.

Seemingly with a drum roll, Linda Potter of Plant City was named the new queen to the shrill sounds of an excited audience of hundreds. Mayor Jack White placed the crown on Queen Linda as the music began again. The evening closed with the

The beautiful Maas Brothers Department Store float parades past the grandstand in the February 19, 1957, Grand Parade.

Queen Linda Potter and her court and cast at the coronation ceremony, 1957.

1957 festival queen and her court leaving the stage in their first recessional as festival royalty.

The ever-popular Baby Parade was Friday, February 22, and once again over one-hundred children in widely diverse costumes proudly marched or rode their little homemade floats past the review stand in Schneider Stadium. The new queen was asked to present the contest winners in the different categories with their loving cups or colorful ribbons. Saturday, the last day of the 1957 festival, was All Hillsborough and Tourist Day and the afternoon brought a popular tourist get-together organized by the chamber of commerce and featured WPLA radio personality Joe Wilson. The night wound down with the awarding of the American Legion new automobile. No longer referred to as a raffle, it was a charity fund activity. So ended the 1957 festival with the closing of the Blue Grass Shows on the midway.

1958

Preceding the 1958 Florida Strawberry Festival was the twelfth annual Junior Agricultural Fair held at the festival grounds and sponsored by the East Hillsborough Chamber of Commerce, with support from state and county agencies as well as the City of Plant City and the Florida Strawberry Festival Association. This three-day fair was an agricultural and homemaking exposition featuring the accomplishments of the 4-H boys and girls, Future Farmers, and Future Homemakers. There was an overlapping of the ag fair officers and directors with the members of the festival association.

The ag fair consisted of a horse show and judging of beef and dairy cattle, hogs, poultry, rabbits, flowers, vegetables, and showmanship. There was also a talent show, tractor driving contests, a frog jumping contest, a feeder calf auction, and the Beef Breeder Banquet. Over time, most of these activities and events would become part of the Florida Strawberry Festival.

The 1958 festival (February 17–22) opened on a cold February day. The Lions Club-sponsored queen selection was moved from the festival grounds to the nearby Plant City High School auditorium, where the swimsuit-clad young girls strolled across the stage as the judges watched carefully. Seven finalists were chosen to move on to the final selection and coronation ceremony on Thursday night.

The Grand Parade followed on Tuesday, with George Carey's informal get-together at the Hotel Plant at 10:30 a.m. and the parade at 1:30. The Grand Parade featured many floats and decorated cars and trucks and among them were the grand floats of the reigning queen and her court and that of the seven finalists. Several high school marching bands rang out John Philip Sousa marches to the delight of the onlookers—young and old alike. The many units passed the review stand at Schneider

Stadium in front of hundreds of spectators and dispersed to join the throngs at the festival grounds.

Wednesday was again Agricultural Day in honor of Agriculture Commissioner Nathan Mayo and all members of agricultural organizations. The Blue Grass Shows were open on the midway until midnight, while exhibits and booths were closed at 10:00 p.m.

The queen final selection and coronation ceremony on Thursday, February 20, was another spectacular evening with pomp and circumstance, royalty and courtiers, music and entertainment. The seven finalists were presented and anxiously awaited the decision of the judges. With a shout of excitement, Sheila West, a sixteen-year-old senior at Plant City High School, was named the 1958 festival queen to be crowned by Mayor Jack White.

The Junior Woman's Club had gained experience in running the Baby Parade and was now the regular sponsor and organizer of this popular and very special event. And Friday, February 21, was another sterling Baby Parade, with approximately one-hundred children aged up to four years in several age group categories. They proudly and excitedly strolled or rolled past the reviewing stand in the stadium with their parents, guardians, or siblings nearby. Queen Sheila was

This dapper youngster enjoys his stroll onstage at the 1958 Baby Parade.

there to present the children with their loving cups and red, white, and blue ribbons.

The tourist get-together on Saturday, the final day of the 1958 festival, was another fun day, again with the leader of the chamber of commerce and radio personality Joe Wilson hosting drawings and awarding prizes. The last major event of the day was the drawing for the new automobile—a beautiful and shiny brand new 1958 Rambler four-door sedan, courtesy of Smith Motors. This was the charity fund event sponsored by the American Legion, who donated the proceeds to a local charity. In the past, the Legion had donated heavily to the South Florida Baptist Hospital drive. Another festival ended with the closing of the midway at midnight.

This brand new Rambler is the American Legion Charity Fund grand prize at the 1958 Florida Strawberry Festival in Plant City.

1959

The end of the decade was a busy year in Plant City. The old 1914 high school building was now housing the Tomlin Junior High, the new city hall was being dedicated, the new county office building was also being dedicated, the large Kwik-Chek on East Baker Street just opened, and Plant City's WPLA radio station celebrated its tenth anniversary. It was also a celebration of Plant City's seventy-fifth anniversary.

Therefore, the twenty-fourth annual Florida Strawberry Festival (March 2–7) became a diamond jubilee celebration commemorating the city's anniversary. The American Legion sponsored the festival, E. N. Dickinson was festival association president and Fred Nulter was the festival general manager. The Lions Club had taken over the queen selection and coronation ceremonies, while the Plant City Junior Woman's Club organized the annual Baby Parade. In honor of the diamond jubilee the festival added an Old Timers' Program and the Jaycees sponsored costume and beard contests. The honored guest at the festival was US Navy World War II veteran Governor LeRoy Collins, who enjoyed having his photo taken while dining on Plant City's famous strawberry shortcake.

The forty-seven queen candidates who stood in the cool night air that Monday for the selection of the final seven were excited, nervous, chilly, and hopeful. The judging went through elimination sessions and while the tellers added up points, there was entertainment for the contestants and spectators. Finally, the judges' seven finalists were selected and announced before hundreds of cheering family and friends. The seven then prepared for the parade and Thursday night's coronation event.

The Florida Strawberry Festival and Plant City Diamond Jubilee Grand Parade had the usual floats and marching bands, and this year added a few antique cars and special dignitaries. Through the downtown streets they paraded and headed for the west side of town and the festival grounds where they passed the reviewing stand and hundreds of local enthusiasts and tourists gathered to take in the thrill of the parade. The Legionnaires, children, and parade participants entered the festival grounds free of charge.

Wednesday, March 4, was Agricultural Day, held in honor of Florida Agriculture Commissioner Nathan Mayo, who had served in that capacity since 1923, and of his cabinet and all others involved in the agricultural industry. The Old Timers' Program started at 2:00 p.m. and prizes

Queen aspirants take the stage for the preliminary selection in 1959.

were offered in seven different categories, including oldest citizen, the farmer planting strawberries for the longest period of time, and other categories.

The Lions Club-directed final selection and coronation ceremony was a well-planned evening and all expectations were met. The stage was draped with a big diamond jubilee strawberry festival sign, and the royal courtiers of heralds, train bearers, flower girls, and evening gown-attired girls celebrated the coronation event. The band played musical selections, there was dancing and singing, and the final announcement of the night congratulated Miss Lynda Eady, who was named the new festival queen. Mayor Otis Andrews proudly did the crowning.

The Friday, March 6, Baby Parade under the direction of the Junior Woman's Club saw over one-hundred entries march in their costumes and ride on their family-constructed floats past the review stand in Schneider Stadium, as anxious parents awaited busy judges. This was always a popular part of the festival and hundreds of family, friends, and parade fans would pack the grandstand for this occasion.

The Woman's Club Flower Show was still a hit and ran all week inside the armory, adjacent to the festival grounds. Saturday brought the final day—All Hillsborough Day and Tourists Day.

A young dancing troupe performs during the 1959 festival queen coronation ceremony.

There was a joyful and fun tourist get-together and prizes. That evening in celebration of the diamond jubilee, the Jaycees sponsored a contest for period costumes and a beard shaving contest for the men, both of which drew a large crowd. The American Legion automobile drawing signaled the final event of the day and the twenty-fourth annual festival ended with the dimming of the lights and the last of the midway rides and shows coming to an end. This, too, was the end of the decade which witnessed some significant changes and many more successful festivals.

THE 1960s: A CHANGING OF THE GUARD, NEW LEADERS, AND NEW IDEAS

The 1960s were years of subtle then dramatic change in the nation. Coming into the decade, people were optimistic. The Eisenhower years had been relatively calm and conservative with an overall growing economy, and the Kennedy years held much promise. Then there was *Sputnik*, technological challenges, Cuba, Elvis Presley, rock 'n roll, folk music, the missile crisis, President Johnson and civil rights and the Great Society, radicalism, the Tet offensive, Vietnam and the anti-war movement, Chicago riots at the Democratic Convention—it was a time of major change. Exiting the 1960s, optimism morphed into a certain negativity, and the country sometimes seemed to be losing its grounding.

Plant City, too, went through many slow but dramatic changes. For the Florida Strawberry Festival Association, it was a change of the old guard—the founders and early organizers of the festival were mostly gone. New faces took their place, and as the earlier citizens had done, they became passionate about the festival and the community that meant so much to them. The Norman McLeod Post No. 26, American Legion, had rejuvenated the dormant festival with a five-day event in 1948 and expanded it to a six-day event in 1950. They slowly relinquished its sponsorship to the festival association itself, with the assistance of the city, Lions Club, chamber of commerce, and Jaycees. Over time the Lions Club and its auxiliary took over the organization of the queen selection and coronation ceremony, the Junior Woman's Club produced the Baby Parade, the Plant City Federation of Garden Clubs presented a changed but continuing annual flower show, and more entertainment was being added to the program.

1960

The silver anniversary of the Florida Strawberry Festival was celebrated February 29 to March 5, 1960. That Monday over fifty booths opened in the morning, and there was a great array of free acts in the early afternoon. At 7:30 p.m. the queen selection process began on the outdoor stage at the festival grounds. Forty-one lovely girls, including the Duyck sisters (a pair of identical twins), paraded across the back-dropped stage before the eyes of the five judges. Singers, dancers, and the Turkey Creek FFA Band entertained the nervous contestants and the excited crowd in the stands. The Lions Club tellers calculated the scores and reported them back to the judges, who then called for another elimination round. At the end of the evening the judges announced the seven finalists to the cheers of hundreds of spectators. The final seven included the Duyck sisters.

Tuesday was not an ordinary day; it began with the dedication of the new city hall at Wheeler and Mahoney Streets in the downtown area. The informal get-together for special guests and dignitaries was held at 11:00 a.m. at the Hotel Plant, just a block away. Ben Rawlins filled in as host in place of former festival president George A. Carey, who had sponsored these events for years. Carey had passed away on July 23, 1959.

The Grand Parade on Tuesday afternoon, March 1, was indeed a grand one. The parade began at 1:00 p.m. and went on for well over an hour, passing by the crowd of onlookers who were thrilled with twenty-two beautifully decorated floats, twelve marching bands, five twirling groups, thirteen Scout troops, thirty-one decorated cars—in all over one-hundred units in the Grand Parade. After passing by the crowd in the stands at Schneider Stadium, the paraders dispersed and entered the festival grounds for a fun day.

Wednesday was a day of free acts and a tourist get-together with fun and prizes. It also brought a change to the festival and it was now themed Agricultural Day and Tourist Day. For years, Agricultural Day had been in honor of Nathan Mayo, Florida's agriculture commissioner. At the age of eighty-three, and fighting lung cancer, Commissioner Mayo did not make the trip to Plant City. He died on April 14, 1960.

The twenty-fifth annual Florida Strawberry Festival Queen Coronation was Thursday, March 3. It opened at 7:30 p.m. under the direction of the Lions Club. Two of the directors of contestants, Barbara Alley Bowden and Ruby Jean Barker Redman, were former festival queens themselves. The master of ceremonies was Lions president and Plant City High School coach and educator Don Yoho. The program was detailed and included the presentation of the 1959 queen and her court and all the courtiers, followed by the presentation of the seven finalists, who were then interviewed by the five judges, none of whom were from Plant City. Dancers and musicians entertained the audience in the packed grandstands while the judges and tellers reviewed the scores. The new court was then presented and the new festival queen for 1960, Betty Jean Cook, was announced. Lynda Eady, the 1959 queen, proudly crowned the wildly excited Queen Betty Jean.

The Baby Parade was the big event for Friday, March 4, and it began at 3:30 p.m. in front of Schneider Stadium. Once again, lines of children formed in widely varied costumes and home-constructed miniaturized floats ready to parade in front of the friendly crowd and the eyes of the judges. Anxious parents and family coaxed them on and awaited the announcements. The winners collected their trophies on the stage by the stadium. The last day of the festival was Saturday, March 5, and the Jaycees had that day. At 3:00 p.m. and again 7:30, the Jaycees presented a series of performers including the young Plant

A group of Shriners enjoys the Tampa Electric pre-parade gathering during the 1960 Florida Strawberry Festival.

Judge James Bruton congratulates Arthur R. Boring, who was named the Outstanding Citizen at a new civic club program during the 1960 festival.

City-born Mel Tillis, along with Webb Pearce, Red Sovine, the Gadabouts and Ernie Lee—all onstage at the festival grounds. The American Legion drawing and awarding of the new automobile was the last event of the evening and of the 1960 festival. After the Blue Grass Shows closed the midway, the festival closed its gates for another year.

1961

The 1961 Florida Strawberry Festival ran March 6 through March 11. It turned out to be a very successful year. The American Legion was still the major sponsor, with the Lions Club organizing the queen pageant. Elmer N. Dickinson was still president of the festival association and Fred Nulter was the general manager. The Blue Grass Shows provided the rides, shows, and the overall midway. There were more than forty booths and exhibits, and the Plant City Federation of Garden Clubs held a flower show all week as did the Plant City Woman's Club, which held a Brandon Area Art Club exhibit. The festival was organized with twenty-one committees and hundreds of volunteers.

The festival opened on Monday with exhibits and booths and at 2:00 p.m. provided several carnival-type free acts. The main event was the first part of the queen selection process which would result in the naming of seven finalists, one of whom would be the next queen with the others serving in the court as maids. In all, fifty-two stunning young girls would parade across the outdoor stage as five judges carefully evaluated their attributes, personalities, poise, and presence. The Lions Club events were detailed and included music and dancing numbers—an entertaining evening for the hundreds of spectators filling the stands. Glenn Knox, the evening's emcee, announced the select seven and the crowds cheered heartily. Next for these girls was the parade and then the Thursday night coronation ceremony.

Tuesday opened with the get-together for special guests and dignitaries at the Hotel Plant hosted by Ben Rawlins followed by the Grand Parade at 12:30 p.m., which began downtown on Haines Street, just west of Evers Street. It was a big parade. There were twenty-nine beautiful and glamorous floats, eleven stirring marching bands, scores of decorated cars, majorettes, twirling girls, Scout troops, and more. The crowds both on the streets and in the stands at Schneider Stadium were in for a real treat.

Wednesday was Agricultural Day again and it was held in honor of Florida Commissioner of Agriculture Doyle Conner, who had just taken

Plant City strawberries began being shipped across the Atlantic in 1961.

office in January. Conner was a farmer by profession, belonged to both the 4-H and FFA, and had served as both the state and national FFA president. Elected in 1960, Conner, who admired former commissioner Nathan Mayo, went on to serve as commissioner of agriculture for the next thirty years.

The selection of the 1961 queen and the coronation ceremonies began at 7:30 p.m. on Thursday, March 9, on the outdoor stage at the festival grounds. It was an elaborate ceremony with the Lions Club Al Berry as the master of ceremonies. With musical accompaniment, 1960 queen, Betty Jean Cook, and seventeen royally attired courtiers made their way across the stage. The finalists were presented to the audience along with the five judges, who would interview the contestants. Following additional entertainment, the 1961 court was presented and the new festival queen, Diane O'Callaghan, were announced to the expectant crowd waiting in the stands. To cheers and applause Queen Diane was ceremoniously crowned by Mayor John Glaros.

The 1961 festival queen and her court promote the East Hillsborough Chamber of Commerce and Plant City strawberries.

The ever-popular Baby Parade was staged Friday at 3:30 p.m. Scores of young children dressed in their exquisite costumes and strolling or riding on their floats paraded by the gathered crowd at Schneider Stadium. The delighted parents and family of the winners in the various categories were presented their loving cups and ribbons onstage. Saturday, March 11, was the final day of the festival and the tourists had their get-together with entertainment and prizes at 3:00 p.m. That evening at eleven o'clock, the American Legion drawing for the new automobile was held and the 1961 Florida Strawberry Festival ended.

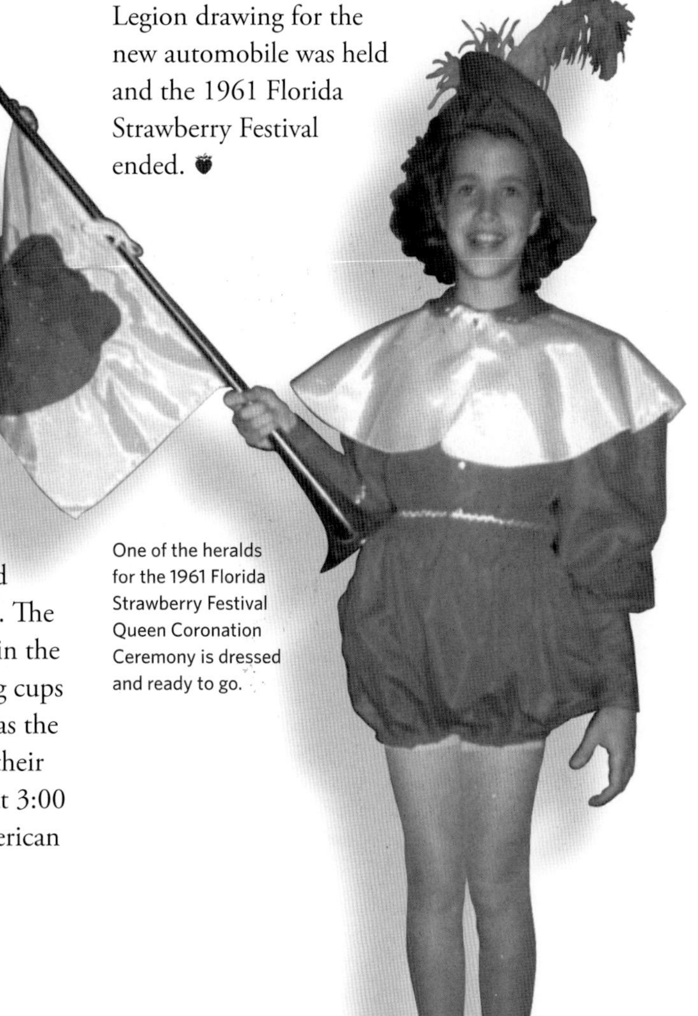

One of the heralds for the 1961 Florida Strawberry Festival Queen Coronation Ceremony is dressed and ready to go.

1962

The year 1962 was a watershed year in some ways for the festival. Prior to the festival itself, there was considerable controversy over whether to expand its scope and include the "fair" concept on a county-wide basis or to remain focused on the local Plant City area and strawberries. Being an official fair would permit the organizers to apply for state and local grants but would require the inclusion of a wider geographic area and more livestock in its exhibits. The festival association decided to try the joint event concept with the 1962 event. This changed the format of the festival for the remainder of its years. A lengthy *Lakeland Ledger* account of the upcoming event read:

> Showing simultaneously with and as an integral part of the annual Strawberry festival, the 15th Annual Hillsborough County Junior Agricultural fair will retain all its features and present 1,500 square feet of agricultural exhibits at what will be known for the first time as the Plant City Strawberry Festival and Hillsborough County Junior Agricultural fair.
>
> Among the events scheduled for the juniors will be the fourth annual frog jumping contest sponsored by the Florida Grower and Rancher magazine on Tuesday evening of the fair, the youth horse show at 7 p.m. on Friday and the Talent show for all young people of school age on Saturday night beginning at 7:30 and sponsored by the Hillsborough County Farm Bureau.
>
> Ten chapters of Future Farmers will be participating with exhibits which, besides strawberry and vegetable, will include a poultry show, dairy and beef cattle exhibits, the cattle exhibit utilizing the space under the grandstand.

Don A. Storms, county supervisor of agricultural education and a member of the Florida Strawberry Festival Association Board of Directors, would oversee the livestock shows. It was an exciting new venture.

E. N. Dickinson remained president of the festival and Fred Nulter was general manager. The American Legion was still the main sponsor, the Lions Club managed the queen pageant, the Plant City Junior Woman's Club arranged the Baby Parade, and the Plant City Federation of Garden Clubs and the Plant City Woman's Club again sponsored their exhibits. This year Tampa Electric joined the special exhibits with its Electrical Exposition. Girls were signing up for the queen contest and Nulter reported that over twenty-five floats and ten bands had already accepted invitations, commercial space in the festival buildings was completely sold out, and all exhibitor space was filled. Another change this year was that the grounds opened all week at 12 noon rather than 9 or 10 a.m.

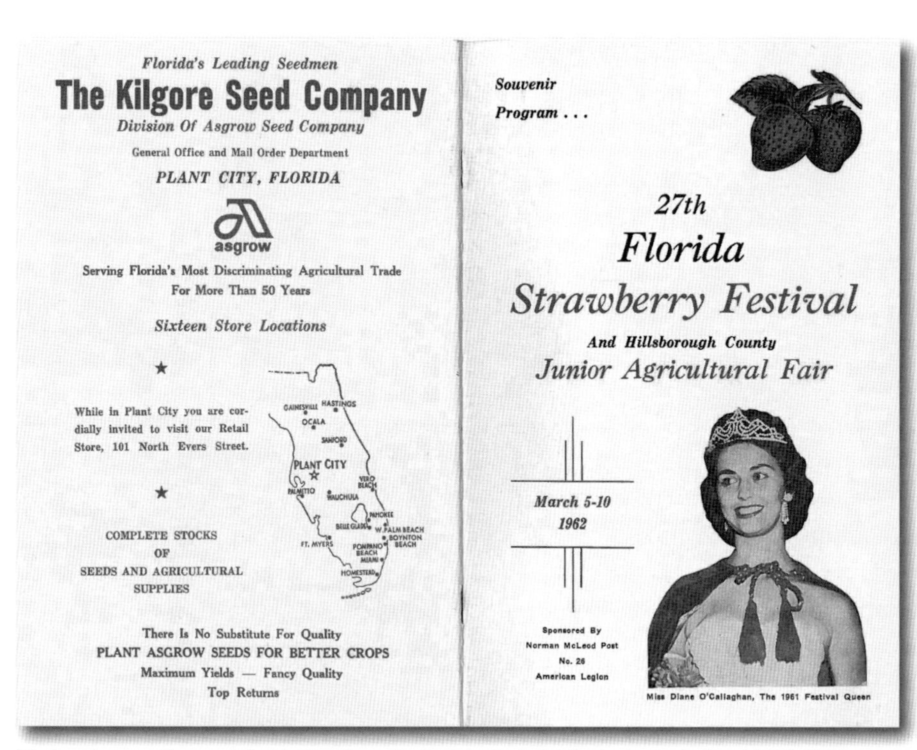
Florida's Leading Seedmen
The Kilgore Seed Company
Division Of Asgrow Seed Company
General Office and Mail Order Department
PLANT CITY, FLORIDA
asgrow
Serving Florida's Most Discriminating Agricultural Trade For More Than 50 Years
Sixteen Store Locations
While in Plant City you are cordially invited to visit our Retail Store, 101 North Evers Street.
COMPLETE STOCKS OF SEEDS AND AGRICULTURAL SUPPLIES
There Is No Substitute For Quality
PLANT ASGROW SEEDS FOR BETTER CROPS
Maximum Yields — Fancy Quality
Top Returns

Souvenir Program . . .
27th
Florida
Strawberry Festival
And Hillsborough County
Junior Agricultural Fair
March 5-10
1962
Sponsored By
Norman McLeod Post
No. 26
American Legion
Miss Diane O'Callaghan, The 1961 Festival Queen

The twenty-seventh annual Florida Strawberry Festival also included the Junior Agricultural Fair in 1962.

Beautifully decorated floats pass by in the 1962 Grand Parade.

The Lions Club reached out to outer communities and girls from Brandon, Dover, and other areas joined the queen contest. The rules for the pageant stipulated these conditions: contestants must be sixteen to twenty-four years of age, unmarried, a resident of east Hillsborough County, and have completed at least her junior year in high school.

The main event on Monday, March 5, was the queen selection which began at 7:30 p.m. Master of Ceremonies Harry McElveen and the five out-of-town judges were ready as fifty-nine anxious young ladies made their way across the stage as they were introduced, one by one. They had earlier modeled specially selected shorts and blouse outfits, and tonight they wore swimsuits. As the judges conferred, a program with the Hungarian Liberty Club, singers, an acrobat, and dancers from Jackie's Studio of Dance entertained the contestants and the expectant crowd in the stands. To the excited cheers of the spectators, McElveen announced the special seven who would be the finalists for the queen pageant the next Thursday night.

The parade day began a bit different also. Tampa Electric had previously been leasing space in the Hotel Plant and in 1961 broke ground for a new building on West Haines Street. The new building became the site for the pre-parade informal get-together hosted by the personable Ben Rawlins, who was the proprietor of the Hotel Plant. The Grand Parade formed on Haines Street not far from the Tampa Electric building and headed through the city streets at 12:30 p.m. Thousands of excited and thrilled locals and tourists alike watched as the flowered and decorated floats, marching bands, and dignitary-filled cars passed by on their way to the stadium and the crowds awaiting there. Another new event that night, added for the ag fair youth, was the "Jumping Frog Contest," which was held onstage at the festival grounds.

Wednesday brought the Baby Parade, usually held on Friday, and approximately one-hundred children aged six months to four years delightedly showed their varied costumes as they paraded past the hundreds of family and friends in the packed stands. The judges announced the winners and

Queen contestants pose picking strawberries in 1962.

distributed the many trophies to the cheers of the crowd.

The usual Saturday tourist get-together event was held on Thursday afternoon with fun, games, entertainment, and prizes. At 7:30 p.m. the beautiful young seven finalists in the queen pageant proceeded onto the ornately decorated stage, with the courtiers and heraldry. After the entertainment program, the judges submitted their scores and the court was announced. Marlene Coon was crowned the 1962 Florida Strawberry Festival Queen and cries, shouts, cheers, and applause broke out. The new festival queen and her court were escorted out to the sounds of celebratory music. This year they were guided to the nearby American Legion Home for a special reception in their honor.

The special event of the Junior Agricultural Fair was the youth horse show reserved for Friday night starting at 7:30. This was a first for the festival. The horse show was competitive, very exciting, and involved seven activities: western pleasure, model Quarter horse, stock seat equitation, open jumping, stock horse, flag race, and cloverleaf barrel race. It was a great evening for the dedicated youthful enthusiasts. Prizes included monetary rewards and rosettes.

The final day of the 1962 festival and Junior Agricultural Fair was Saturday, March 10, and had additional changes to the previous programs. It was Farmers' Day and the chamber of commerce sponsored a special fish fry for the farmers, after which Commissioner of Agriculture Doyle Conner addressed the attendees. This was followed by the youth talent show held onstage and the many singers, dancers, and musicians were judged by local celebrities. The last event was the American Legion-sponsored drawing for a new automobile and a lucky ticket holder drove home happy. As the Blue Grass Shows midway rides wound down, the festival drew to a close.

1963

Although the focus tends to be on parades and queens, the many exhibits and booths open for virtually the entire festival draw thousands of visitors each year. The 1963 festival was a great example because this was the first Hillsborough County Fair and twenty-eighth Florida Strawberry Festival with expanded livestock and other fair-associated activities of the overall exposition. The American Legion Post and the Lions Club received dual billing as sponsors and aided the reconstituted festival association in organizing the entire festival. E. N. Dickinson continued as president of the festival association and G. R. Patten took the reins as the festival general manager.

The 1962 experiment of joining a fair with the festival delighted many and concerned others. Nonetheless, it was decidedly a success and would continue through 1991, with numerous adjustments along the way. Here is the foreword printed in the 1963 program:

> Plant City and East Hillsborough County has long been noted for the excellence of the strawberries raised here. This reputation has been maintained down through the years and constitutes a valuable asset to the growers of this particular section of Florida. Plant City and Dover, five miles west, are the marketing points for this area.
>
> Other crops and industries such as spring and fall vegetables, citrus groves, canning

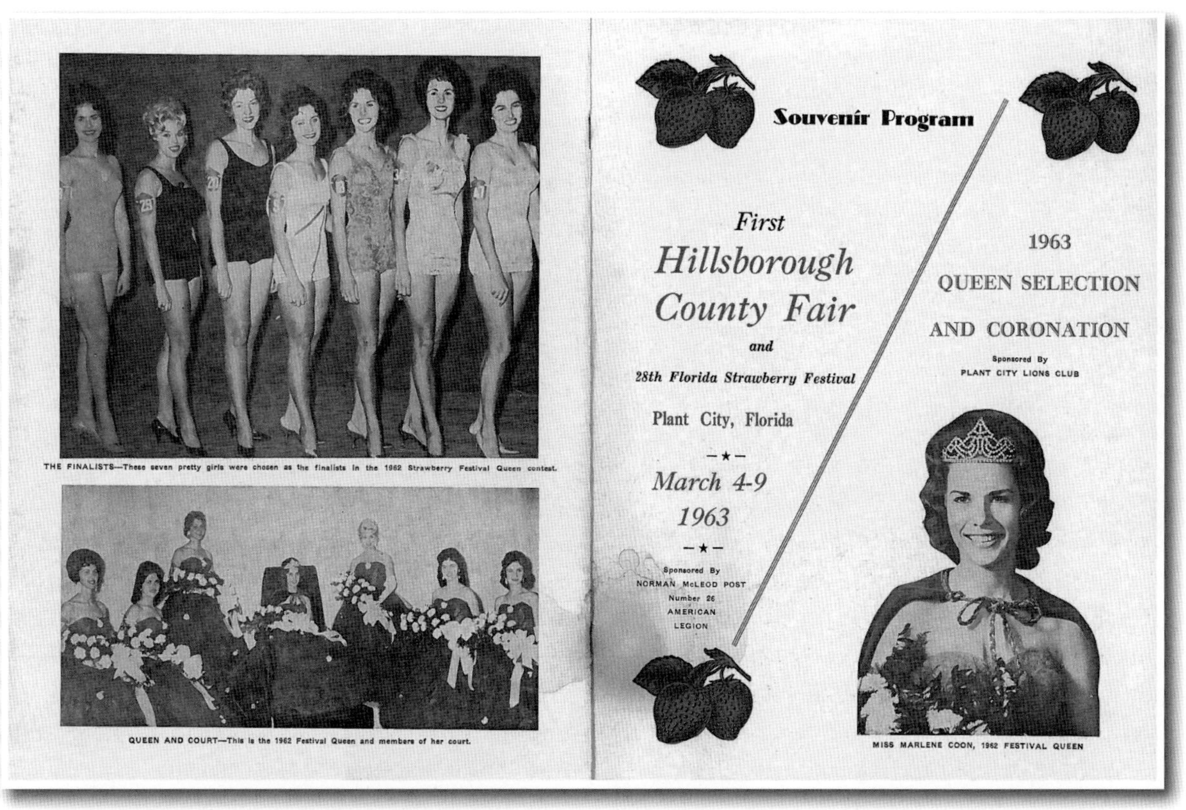

The First Hillsborough County Fair is coupled with the twenty-eighth Florida Strawberry Festival in March 1963.

plants, cattle, timber, and phosphate have contributed greatly to the progress and prosperity of Plant City and to the entire area of Hillsborough County.

Mindful of this diversification, the Board of Directors of the Strawberry Festival voted in 1962 to change the name of this festival occasion to the Hillsborough County Fair and Florida Strawberry Festival.

The Board of Directors of the Hillsborough County Fair and Florida Strawberry Festival presents this annual exposition with the sincere hope that it may prove worthy of its predecessors and a pleasure to all who view it.

A week before the 1963 festival was to open, tornado-like winds, up to seventy-five miles per hour, ripped the Plant City area causing much damage. Opening day, Monday, March 4, was cool but not stormy. That night fifty-three young ladies vied for the crown and ascended the stage at the festival grounds as Master of Ceremonies Hilman Bowden introduced them to the five judges. While musicians, several singing groups, a gospel quartet, square dancers, and Jackie's Dance Studio students provided entertainment, the judges evaluated each of the aspiring girls. Bowden announced the decision of the judges and the seven gleeful young ladies were introduced, this time as the finalists in the competition for the title of festival queen.

Once again, the Plant City Woman's Club sponsored a Brandon Area Art Club exhibit, which ran all week. The Plant City Garden Club sponsored their annual flower show in the armory on Thursday and Friday.

The informal get-together before the parade was now called social hour and special guests and many dignitaries, including the commissioner of agriculture and the mayor of Tampa, gathered at the Tampa Electric building on Haines Street at 10 a.m. The Grand Parade, held on Tuesday, March 5, formed near the Tampa Electric building and proceeded through downtown, raising the spirits of the thousands of onlookers as over one-

During the 1963 festival, Edgar Hull was named the Outstanding Citizen at the joint civic club luncheon.

hundred ornately decorated floats, bands, military, other marching units, and cars full of dignitaries paraded the five-mile route to the stadium and before thousands of cheering fans packing the grandstand.

Wednesday was a day for the livestock. At 7:30 p.m. stock judging was presented in front of the grandstand, with ribbons and prizes for the many winners. Then came Thursday's popular event—the final selection and coronation of the queen. Jack Dempsey, an association director and master of ceremonies, introduced the program and the judges, followed by the presentation of reigning Queen Marlene Coon and her court. Four former festival queens entertained the spectators and the seven finalists were introduced and prepared for the judges' interviews. After further entertainment, the emcee revealed the members of the new court and announced sixteen-year-old Janice Barnes as the 1963 festival queen. The *Lakeland Ledger* reported (Friday, March 8, 1963), "Tears of happiness flowed as the 16-year-old Turkey Creek High School senior was crowned queen of the festival by Mayor Arden Mays. They continued to flow moments later as she got a warm congratulatory hug from her parents, Mr. and Mrs. P. H. Barnes." The festival queen and her court, to the resounding cheers of the happy crowd, proceeded off the stage to the American Legion Home for a special reception with strawberry punch.

The very special children's event, the Baby Parade, was held on Friday, March 8. Once again close to one-hundred youngsters, some giddy and some fearful, paraded their finest costumes and small floats across the front of the stadium to the stage, escorted by parents or family members, as they passed the judges panel. The categories were in age groups: six months to one year, one to two years, and three to four years. The prizes were for costumes, floats, healthiest looking, and personality. At the end a king and queen of the Baby Parade were crowned and photographed—usually with smiles and grins but occasionally with a tearful expression. Parents were enthusiastic about the prize ribbons and trophies.

Saturday, March 9, was All Hillsborough and Tourist Day and events included the tourist get-together with its fun, entertainment, and delightful prizes for farthest traveled, oldest, etc. With the 11 p.m. American Legion new automobile drawing making another festival patron happy, the lights went off at another festival, this time the first Hillsborough County Fair and Florida Strawberry Festival combination.

This nautical entry adds variety to the 1963 Grand Parade.

1964

The second Hillsborough County Fair and twenty-ninth Florida Strawberry Festival got off to a different start. It did not have a permit. An Agriculture Commission inspector on Sunday had "ruled out a 'digger' game in the midway and ruled off the premises a machine with a film strip of a naked woman. Police Chief Bob Spooner said . . . the permit would be forthcoming when these and other minor corrections were made." (*Tampa Tribune*, Tuesday, March 3, 1964). Festival President E. N. Dickinson, who was also the commander of Norman McLeod Post 26, American Legion, and G. R. Patten, festival general manager, along with the Lions Club were ready and opened at 12:00 noon on Monday, March 2. The Plant City Woman's Club-sponsored Brandon Area Art Club Exhibit was in place and set to run all week.

Monday night the selection began for the queen finalists and this night, with ideal weather, one-thousand fans crowded the grandstands to watch sixty-eight young female candidates promenade across the stage. Introduced by Master of Ceremonies Cecil Everidge, they passed the judging station with poise. The entertainment included a musical combo, dance routines, and a Cordovox (similar to an accordion), while the judges concluded their evaluations. The program ran a bit long but when the announcement of the final seven was made, cheers went up from the stands and girls let out screams of joy, while others, though not chosen, joined in the celebration.

Tuesday, March 3, was parade day and the "Social Hour" for politicians, dignitaries, and special guests was sponsored by Tampa Electric and held at its downtown Plant City headquarters. Hosted by Ben Rawlins, this year it included a sumptuous lunch buffet and local famous strawberry shortcake. At 12:30 p.m. with festivity in the air, the Grand Parade began with "a record number of floats and 75 units in all. Gov. Farris is slated to lead the parade and politicians from all over the state will be here." (*Tampa Tribune*, Tuesday, March 3, 1964). The parade went very well and throngs of people, both downtown and in the stadium grandstands, were thrilled with the sight and sounds of this large festival parade.

Exhibitors Day was the theme for Wednesday, March 4, and the only special event was the Lions, Kiwanis, and Rotary Club lunch meeting

The Tampa Electric Company float makes a splendid show in the 1964 Grand Parade en route to the festival grounds.

at the Hotel Plant where the "Outstanding Citizen" would be honored. The event was initially sponsored by the First National Bank in Plant City in 1960 and became a joint civic club program. This day Willard D. McGinnes was named Outstanding Citizen for 1964. Previous honorees were Arthur R. Boring, 1960; Quintilla Geer Bruton, 1961; Frank H. Moody, 1962; and Edgar Hull, 1963.

Exhibits, booths, and the midway were open at noon on Thursday and by the night of March 5, the excitement in the air was palpable. At 7:30 p.m., following a musical prelude, Horace Andrews of Badcock Furniture, who was the emcee, introduced the judges and presented the 1963 queen and court. Draped in royal attire, the heralds, crown bearer, maids, flower girls, train bearers, and the reigning queen made their entrance. Shortly after, the seven finalists were presented and escorted onstage to meet the judges. Musicians and dancers entertained the crowd while the judges conferred; with the final decision made, the emcee announced the court, then the queen. Georgie Blevins, a student at Chamberlain High School, became the 1964 festival queen. Mayor Arden Mays smiled as he placed the crown on Queen Georgie, who was still beaming after hearing her name called out.

For weeks, parents and their children had been waiting for this moment and on Friday, March 6, the Baby Parade began at 3:30 p.m. in front of the grandstands at Schneider Stadium. The Plant City Junior Woman's Club organized the parade and over one-hundred children in various age groups and in original and creative costumes and floats passed by the reviewing stand for their judging. Onstage the proud parents and children received their trophies and ribbons.

A different and novel event was added to the Friday program and a judo exhibition, directed by the Plant City Recreation Department, was staged in the stadium beginning at 7:30 p.m. And, again, the Plant City Garden Club held its flower show on Friday and Saturday from 1:00 to 8:00 p.m. in the armory building. It always drew a large crowd for its spectacular and colorful displays and arrangements.

Saturday, March 7, was the final day of the exposition and a unique event was scheduled for the early afternoon. The pet show opened at 1:00 p.m. for judging of miscellaneous pets, and at 1:45 the pets paraded before the grandstand and scores of amused and entertained spectators cheered them on. At two o'clock the obedience pet judging took place, followed by a trained dog exhibition. It was also Tourist Day and the tourists who registered by 3:00 p.m. enjoyed their get-together, fun, entertainment, and prizes, all sponsored by the East Hillsborough Chamber of Commerce. The Blue Grass Shows, rides, and acts on the midway shut down at midnight; the tents began coming down and the 1964 festival wrapped up another successful year.

Newly crowned 1964 Strawberry Festival Queen Georgie Blevins and her court prepare for photographs onstage.

1965

The third Hillsborough County Fair and thirtieth Florida Strawberry Festival ran into some cold weather on March 8 through March 13, 1965. Organizers remained the same except C. D. Swingley took over as general manager of the festival. The organizers, the entire board of directors of the festival association, the members of the Lions Club, and the hundreds of others who made these festivals possible served as volunteers, drawing no financial compensation. In a slight change from last year, the gates to the fairgrounds opened daily at 11:00 a.m.

The thirtieth Florida Strawberry Festival makes its appearance with the third Hillsborough County Fair in 1965.

The Monday of opening day was chilly and when the sun dropped it became cold. As the sixty-seven contestants emerged onto the stage, being introduced by Master of Ceremonies Jim Redman, they were shivering in their swimsuits. They quickly promenaded past the careful eyes of the judges, then awaited the results. The entertainment kept things lively for the young ladies and the hundreds of fans in the grandstand with dancing, singing, a fashion show by the Hillsborough County Homemakers Council, a pantomime, baton twirling, a harmonica solo, and the Zephyrhills FFA Band. Redman announced the sweet seven finalists in the queen contest and the contestants cheered, the crowd applauded, and all moved quickly to find warmth.

A *Plant City Courier* article (Thursday, March 4, 1965) reported the upcoming fair and festival under the headline: "Parade Will Be Largest Ever Staged:" "Glittering floats, pretty girls and plenty of peppy bands will highlight the gala Hillsborough County Fair and Festival

The seven finalists shiver on the outdoor stage in the 1965 festival queen contest.

parade Tuesday as giant spectacle sparkles through the downtown area and passes in review at the Fair grandstand . . . Heading the procession will be Gov. Haydon Burns, members of his cabinet, plus other state and county officials . . . More than 100 units have entered the parade—the largest number in the history of the event. Twelve bands will also participate."

Queen Sandra Link and her court wave to the crowd during Grand Parade.

This stunning float gets the attention of the crowds watching the festival's 1965 Grand Parade on a beautiful day.

After the social hour at the Tampa Electric office building, the Grand Parade stepped off at 12:30 p.m. and headed through the city streets onto Reynolds Street and to the fairgrounds to pass by the review stand and the stadium grandstands where hundreds more fans cheered them on. The seven finalists rode in a new convertible, hatless but clad in warm jackets and coats.

Wednesday, March 10, was Merchants Day and focused on the booths, exhibits, and midway acts and shows. At the Kiwanis and joint civic club luncheon, Alex B. Hull III was honored with the Outstanding Citizen Award.

Thursday was the big event—the final selection and coronation of the new queen. The price of admission had not changed and remained at thirty-five cents. At 7:30 p.m. the program began with a musical prelude followed by Master of Ceremonies Al Berry's introduction and presentation of reigning Queen Georgie Blevins and her court, with ceremoniously attired courtiers accompanying them. Berry introduced the seven finalists who then met the judges for interviews. Following an entertainment program, the judges submitted their decision and the new queen and festival court were introduced. Young Sandra Link was now the 1965 Florida Strawberry Festival Queen. She smiled happily as she accepted the crown. Among the prizes for the queen was a trip to Nassau via P & O Steamship Lines and a weekend at the Beachcomber on Miami Beach for the festival queen and court.

The annual Baby Parade started at 3:30 p.m. on Friday, March 12. The Junior Woman's Club organized the one-hundred entrants into age groups, and they paraded across the area in front of the grandstand to the platform in review of the judging panel. The children wore intricately constructed costumes and some rode on detailed mini-floats. They were adorable. The parents, families, and friends were nervous, anxious, and having a wonderful time. They reveled in the presentation of the winners' trophies and ribbons, distributed onstage.

There were two interesting additions to the Friday program. At 8 p.m. there was a judo exhibition in front of the grandstand, followed an hour later by the Strawberry Strollers Square Dancers who put on an exhibition in the center of the main exhibit building. The beautiful Plant City Garden Club's annual flower show was staged in the armory building and was open from 1 to 9 p.m. on both Friday and Saturday nights.

Prior to opening the gates to the fairground on Saturday, March 13, in another new addition to the program, the Plant City Horseshoe Pitching Club sponsored an open horseshoe pitching contest starting at 10 a.m. At 1 p.m. the pet show opened and at 1:45 the Parade of Pets in front of the grandstand began, with numerous dogs and other miscellaneous pets led before the appreciative pet-loving crowd. The flower show closed at 9 p.m., the exhibits that Saturday closed at 10:30, and the Blue Grass Shows midway rides and acts closed at midnight and another fair and festival ended.

1966

The year 1966 brought the Hillsborough County Fair and Florida Strawberry Festival running March 7 through 12—this time the Norman McLeod Post No. 26, American Legion, was not listed as the sponsor. In its place was the Plant City Lions Club. C. D. Swingley was the general manager and Elmer N. Dickinson was the president and chairman of the festival association. Booths and exhibit space were sold out—the Plant City Woman's Club sponsored the Brandon Area Art Club Exhibit all week, and the Plant City Garden Club sponsored its annual flower show on Friday and Saturday from 1 to 9 p.m. In addition, tourists could register daily for drawings for free Plant City-area strawberries.

The 1966 fair and festival opened Monday, March 7 at 11 a.m. and people were preparing for the queen finalists selection set for 7:30 p.m. When it started, the numerous aspiring young ladies in swimsuits were introduced by emcee Al Berry. They sashayed from the platform down the lit runway extending out into the audience, then back, passing the judging panel. An entertainment program of song and dance helped relieve the tension of the moment while the judges concluded their selections. Berry then announced the successful seven: Cheryl Yvonne Barrington, Pamela Chambers, Janie Louise Goodell, Sandra Kay Leitner, Alice Kaye Mabry, Kathy Mascali, and Linda Schweitzer—one of whom would be chosen festival queen at the pageant Thursday night.

After the prestigious morning social hour on Tuesday, March 8, at the Tampa Electric office building downtown (hosted by Ben Rawlins, whose Hotel Plant was demolished in 1966 and replaced by the Hillsboro Bank), the Grand Parade began. C. D. Swingley was quoted to have said, "This year's parade will be the biggest ever and will have over 100 entries, including business floats, bands and agricultural floats." It was big. Thousands of spectators, local and tourists alike, delighted in the festive parade atmosphere. The paraders finally reached the stadium, passed the

Local businessmen compete in the exciting strawberry picking contest during the 1966 festival.

review stand and the cheering audience, and dispersed, many to enter the fairgrounds to enjoy the event. Legionnaires, children, and parade participants were granted free admission.

In addition to Wednesday, March 9, being Merchants Day whereby patrons could pick up free coupons at Plant City stores and use them for half-price admission and midway rides, Wednesday was also included a new event—the cook-off in the strawberry recipe contest, set for 2 to 4 p.m. in the Electrical Exhibition Building. Eight finalists in the 1966 First National Bank recipe contest would prepare their recipes for judging. The first-place prize was an electric range! The seven other finalists received $25 savings bonds. And at the joint civic club luncheon, Robert Trinkle was named Outstanding Citizen.

The coronation was the main event for Thursday, March 10, and the weather was cooperative for the outdoor program. After the musical prelude at 7:30 p.m., emcee Al Berry welcomed the audience, introduced the five judges, and presented the 1965 queen and court along with their courtiers. This was followed by a similar entrance by the seven finalists, who were then introduced to the judges for interviews. The entertainment interlude included Jacquelyn's Dance Studio dancers, the Warsaw Concert with David Turner on the piano, and soloist Marche Crocker accompanied by Turner. Berry then presented the 1966 court and courtiers, not in trappings of royalty but in tuxes and gowns, and announced Kathy Mascali queen of the 1966 Florida Strawberry Festival. She was overwhelmed and received her crown from a smiling Sandra Link, the now former queen. This year the prizes for the queen and her escort included a trip to Nassau on the gracious steamship *SS Florida* and a weekend for the queen, court, and escorts at the Suez Resort Hotel on Miami Beach. Prizes were sponsored by radio station WPLA, the Lions Club, and the festival association.

Baby Parade March 1966
Christa Kantrowitz & her mother Joyce
Christa won the "Most Beautiful Girl" award

Judged "Most Beautiful Girl" in the 1966 Florida Strawberry Festival Baby Parade, this youngster waves to the crowd.

Friday at 3:30 p.m. was the eagerly awaited Baby Parade with close to one-hundred children dressed in a panoply of costumes, including celebrity themes, cowboys, princesses, and farmers. Accompanied by their parents, they marched in front of the grandstand to the applause of the hundreds of onlookers to the reviewing stand and their trophies and ribbons. There was also a horseshoe pitching contest, a national strawberry shortcake eating contest, and a judo exhibition, which seemed like a good combination.

Saturday, March 12, was the last day of the combined fair and festival and included an all-day horseshoe pitching contest, a square dancing exhibition, a dog obedience show, a demonstration of dog handling by the Hillsborough Sheriff's Office, and the continuation of the booths, exhibits, and Blue Grass Shows' midway rides and acts until the lights went off at midnight. Another festival closed its tents.

1967

As with most every year, there was a lot going on in 1967. Nationally, Vietnam was still a hot war and a controversial topic. The *Apollo 1* tragedy shook the nation. At home in Plant City the population was stuck on no growth (15,400), while Hillsborough County continued to grow (490,000). Plant City did have its new shopping center, the Plant City Plaza (also called Grants Mall) and Hillsborough Community College was showing new life in the community.

At the festival, things were looking good. E. N. Dickinson was president, Jack Dempsey was vice president, C. D. Swingley was general manager, and the Lions Club, Junior Woman's Club, the Plant City Woman's Club, and the Garden Club were in complete support. The opening time was changed back to 12 noon, the exhibits would be open until 10:30 p.m., the Blue Grass Shows' midway rides and acts would close at midnight, and the art show and a new entry—the Plant City Builders Model Homes—would exhibit all week. Tourists could register each day for the free strawberry prizes. Last, the use of the space under the stadium grandstands was working well for the cattle and farm equipment exhibit.

On Monday, March 13, the fifth Hillsborough County Fair and thirty-second Florida Strawberry Festival opened with a two-day horseshoe pitching tournament. The main event was the queen selection, which drew fifty-four contestants. Although this event began at 7:30 p.m., it began earlier for the young ladies. There was the registration process, followed by coaching and rehearsals through Sunday. Monday, they were ready and as they were introduced by emcee John Ryals, they promenaded down the runway to the platform stage in review of the five judges. After the entertainment portion of the program, allowing time for the judging, the seven fortunate girls were announced and called forward for celebratory photographs.

The big parade day followed on Tuesday, March 14, and the social hour at the Tampa Electric office building began at 10 a.m., with Ben Rawlins welcoming dignitaries, civic leaders, and special guests to the mingling and luncheon event. Governor Claude Kirk was among the guests and joined the Grand Parade immediately after the luncheon. Again this year, it was "Hailed As Biggest And Best In The Events' History," with over one-hundred units, floats, marching bands, marching groups of all sorts—and brilliant weather. They paraded through the city streets and past the cheering crowds along the route and in the stadium before dispersing.

Wednesday, March 15, was Merchants' Day, offering half-price admission and rides with merchants' coupons. It was also the day for the joint civic club luncheon sponsored by the Kiwanis Club at Johnson's Restaurant. This day former mayor and city commissioner Bill Rickert was the honoree and Robert Trinkle, the previous recipient, introduced Rickert as the 1967 Outstanding

Tourists and visitors relax at the Winter Visitor Club during the 1967 festival.

Citizen. The last special event of the day was the cook-off in the strawberry recipe contest sponsored by First National Bank and held at the Electrical Exhibition Building from 2 to 4 p.m. First prize again this year was an electric range.

The ornate and decorous selection of the queen and coronation ceremonies began at 7:30 p.m. on Thursday, March 16, at the outdoor stage at the fairgrounds. Again, Al Berry served as master of ceremonies and introduced the judges, presented the reigning queen, Kathy Mascali, and her court, and introduced the seven finalists for queen as they sashayed along the red carpeted runway to meet the judges. After musical numbers and dancing entertained the hundreds of enthusiastic fans in the grandstand, Berry called out the names of the festival queen's court and announced the new queen, young Maria Junquera, who swooned and smiled widely as she received the crown from Mayor Dick Elston. The new festival queen, her court, and nine courtiers (including two heralds, a ring bearer, two train bearers, and four flower girls) posed for numerous photographs and congratulations from family, friends, and an adoring audience. The queen and her court were then special guests at an American Legion and Lions Club reception at the Legion home nearby.

Smiling Plant City Mayor Dick Elston places the ornate crown on Queen Maria Junquera at the 1967 Florida Strawberry Festival.

The Junior Woman's Club organized the annual Baby Parade slated to be the largest ever. By Friday, March 17, over two-hundred children had signed up at the Lad & Lassie shop and families were preparing with glee. *Courier* editor Kathryn Cooke said the club had transformed the outdoor stage "into a fabulous setting for the 1967 annual Baby Parade on Friday afternoon." The children would be "clad in a variety of costumes" and would "jet into the hearts of spectators as the youngsters cross the stage drawing rounds of applause." She continued, "Climaxing the colorful event will be the crowning of a king and queen from the four-year-old group." A court was also selected, and the parents, family, and friends delightedly photographed the smiling children and proudly accepted the trophies and ribbons.

Saturday, March 18, was the last day of the festival and was themed All Hillsborough and Tourist Day. Besides the booths and exhibits, it featured the World's Champion Strawberry Shortcake Eating Contest at 8 p.m. Following was a program by the Plant City Recreation Department including square dancing and "varied entertainment." Another successful exposition ended when the Blue Grass Shows' midway shut down at midnight.

1968

Things were looking good for the joint fair and festival in 1968. Just as last year, E. N. Dickinson was president, Jack Dempsey was vice president, C. D. Swingley was general manager, and the Lions Club, Junior Woman's Club, the Plant City Woman's Club, and the Garden Club were in complete support. The gate opening remained at 12 noon, the exhibits would be open until 10:30 p.m., the Blue Grass Shows' midway rides and acts would close at midnight, and the art show, the Plant City Builders Model Homes, and this year the flower show would exhibit each day of the festival. The tourists could again register each day at the chamber of commerce booth for the free strawberry prizes. Of the 145 out-of-state visitors from seventeen states and Canada who registered on Monday, ten won pints of strawberries.

The sixth Hillsborough County Fair and thirty-third Florida Strawberry Festival opened on Monday, March 4, and again the booths and exhibit spaces were filled. The main event that evening would be the selection of the seven finalists in the queen pageant which began at 7:30. A line of sixty-three hopeful girls stepped onto the outdoor stage at the fairgrounds. As they were announced, they passed by the judging panel and took their places. The Lions Club had again organized the selection program, and the music and dancing following the judges' review was highly entertaining to the gathered audience of family, friends, fans, locals, and tourists as they awaited the results. With full voice, Master of Ceremonies Bob Spooner, who was also the Lions Club president, Plant City Police chief, and Legionnaire, announced the names of the successful seven young ladies who would compete for the title of queen on Thursday night.

Tuesday, March 5, was a great day for a parade. By the end the social hour, the pre-parade social event and luncheon for dignitaries and special guests held at the Tampa Electric office building in downtown Plant City, the temperature rose to seventy degrees. The hundred or so floats, bands, marching units, cars, and other participants formed on South Evers Street near the Plant City Plaza and marched up Evers, over on Haines, up Collins, then west on Reynolds to the festival grounds before a record crowd of thousands. The two-and-a-half-hour Grand Parade "was the largest ever." In a strange coincidence, a parked car blocked the railroad tracks. An article in the *Courier* read, "Passengers on the Seaboard Coast Line's crack train, the Champion, had a chance to take a long look Tuesday at some of the parade floats on West Haines Street." The conductor, Bill Herold, and Horace Hancock managed to push the car far enough to clear the tracks. "Meanwhile, the train's passengers had a grandstand seat for the parade as it waited to begin on nearby Haines Street."

The next day, Wednesday, March 6, was themed Merchants' Day and Civic Club Day. The Kiwanis-sponsored joint civic club luncheon

The seven queen finalists greet cheering spectators during the 1968 Grand Parade.

was held at noon, and Plant City's "Goodwill Ambassador" and host for the many pre-parade luncheons, Ben Rawlins, was introduced by Bill Rickert as the Outstanding Citizen of 1968. Rawlins was the ninth person to be so honored. Being Merchants' Day, anyone attending the festival could get half-price admission and rides with coupons available for free at merchants in Plant City.

Thursday, March 7, was the big day for the final queen selection and coronation, but the weather was not promising. The program began with Lions Club director and WPLA radio announcer Al Berry welcoming the audience and introducing the five judges, one of whom was the popular Tampa mayor, Dick Greco. Then came the presentation of the reigning queen, Maria Junquera, and her court, along with the heralds, crown bearer, flower girls, and trainbearers. Berry then introduced the seven finalists, who met with the judges for their interviews. The shortened musical interlude entertained the anxious and suspenseful audience while the judges met. The emcee rose to introduce the new court and announced Silvia Azorin as the Florida Strawberry Festival Queen for 1968. A proud Mayor Dick Elston placed the crown on a beaming Queen Silvia.

The horseshoe pitching tournament opened early on Friday, March 8—Baby Parade Day. The parade, with 162 children ready to strut their stuff, was anticipated in this Thursday article in the *Courier*: "The wee ones will coo and cavort their way straight into the hearts of spectators Friday when the annual Baby Parade is staged at the Fair and Festival grounds Friday afternoon. A total of 162 youngsters are entered in the 3:30 p.m. event sponsored by the Junior Woman's Club . . . Climax of the production will be the crowning of a king and queen from the four-year-old group. Many of the tiny contestants will be riding on elaborate floats; some will toddle along under their own power; and others are carried in the arms of their mothers." It was a long and fun-filled parade with the winners winding up on the stage with their parents to accept their awards, trophies, and ribbons. The final event Friday night was the eight o'clock Free Gospel Sing at the festival grounds.

Saturday's theme was All Hillsborough County and Tourist Day. In addition to all the booths and exhibits, midway rides and shows, the Plant City Horseshoe Pitching Club sponsored a second day of the tournament. At the close of the night, the Blue Grass Shows' rides and acts went dark and the 1968 fair and festival ended.

Queen Silvia Azorin and her court adorn the stage at the 1968 coronation ceremony.

1969

The program for 1969 simply read, "Plant City Strawberry Festival and Hillsborough County Fair, Plant City, Florida." The fair and festival ran from March 3 through March 8. It was the last of the fourth decade of strawberry festivals.

The festival association was operating well and the Lions Club, Junior Woman's Club, the Plant City Woman's Club, and the Garden Club continued to be in full support. The gate opening remained at noon, the exhibits would be open until 10:30 p.m., and the Blue Grass Shows' rides and acts would close at midnight. The art show, the flower show, and the new hobby show (sponsored by the Business and Professional Woman's Club) would all exhibit each day of the festival. Tourists could again register each day at the chamber of commerce booth for the free strawberry prizes. An item of interest was the winner of the shortcake eating contest—Larry Gregory ate six pounds and four ounces of shortcake. The Parkesdale Farm Market on Highway 92 was also now open for the 1969 festival.

Interestingly enough, the 1969 souvenir program simply lists the event as the Plant City Strawberry Festival and Hillsborough County Fair.

On Monday, March 3, the fair and festival opened at noon to the usual large opening-day crowd and again the booths and exhibit spaces were sold out. That night the queen selection event began at 7:30 and Lions Club Master of Ceremonies Vernon Ayscue introduced the lovely young ladies as they entered the stage area and passed by the panel of judges. Following the entertainment program of dancing and song, Ayscue called the names of Sheila Diane Howard, Robyn Garrels, Yvonne Almon, Margie Tew, Dee Zoet, Judy Gay, and Kathy Hockett—the seven

The FFA, shown here circa 1969, officially took part in the livestock shows after the Hillsborough County Fair joined with the Florida Strawberry Festival.

finalists. Hundreds of pageant enthusiasts, families, and friends broke into applause. The seven excited ladies celebrated and began thinking about the big parade and the coronation ceremony.

The Grand Parade started on Tuesday, March 4, following the social hour and luncheon at the Tampa Electric building on Haines Street. The parade formed by the Plant City Plaza and traveled north on Evers Street to Haines Street, then Collins Street, Reynolds Street, and on to the festival grounds, passing before the grandstand in review of the judges and thousands of festival revelers gathered for the wonderful occasion. Legionnaires, children, and parade participants were admitted free to the grounds. And, something new that year, the Kiwanis Club sponsored a talent show at five o'clock that evening.

Merchants' Day and the civic club luncheon both were on Wednesday, March 5, with the Kiwanis-sponsored luncheon being held at noon at Johnson's Restaurant on Haines Street. Horace H. Hancock, publisher of the *Courier*, was introduced by Ben Rawlins and named the Outstanding Citizen for the year. At the festival, patrons could get half-price admission and rides with merchants' coupons available free in the stores. At 8 p.m. was another new event—the fashion show sponsored by Beeline Fashions, an organization that contracted with hostesses and "stylists" to put on home parties to show and purchase women's clothing.

The main event on Thursday, March 6, was the selection and coronation of the Florida Strawberry Festival Queen. Because of the weather, the event was moved to the Plant City High School auditorium on Woodrow Wilson Street. Hilman Bowden was the master of ceremonies and welcomed the hundreds in attendance and introduced the five judges. The 1968 queen, Silvia Azorin, and her court and seven courtiers were presented and promenaded to the stage, after which Bowden introduced the seven hopefuls to the crown. A brief entertainment program by the Tempos, with piano accompaniment by Rosemary Buckingham, was then followed by the presentation of the Miss Congeniality Award by Tom Barwick. Finally, the court members were introduced and the new queen, Dee Zoet, was called and emotionally received her crown, as her maids and the delighted audience applauded and cheered. It was the last coronation of the decade.

A horseshoe pitching tournament started early on Friday, March 7, and the main event, the Baby Parade, was set for 3:30 p.m. Well over one-hundred children were registered and the Junior Woman's Club organized another wonderful parade at the stadium in front of the grandstand. The babies walked and rode, in original and widely varied costumes and floats, passing the judges and the happy family, friends, and ebullient crowd on their way

These seven finalists prepare for the final queen selection at the 1969 Florida Strawberry Festival.

This 1969 queen coronation ceremony features Queen Dee Zoet and her court onstage at the festival grounds.

to the stage area for final review and, hopefully, prizes of loving cups and ribbons. Then there was the crowning of the little king and queen of the Baby Parade. It was, as usual, adorable. The evening wrapped up at the fairgrounds with the seven o'clock Homemakers Extension Service Fashion Show and the Kiwanis-sponsored talent show, both on the stage before the grandstand.

Saturday, March 8, started with a final horseshoe pitching tournament and the special events ended with the eight o'clock Free Gospel Sing, which included four different groups of singers at the grandstand. The Blue Grass Shows, with its midway rides and acts, ended at midnight, and the last night of the 1969 Florida Strawberry Festival was over.

Endnotes

1. Johnny J. Jones, whose exposition shows filled the midway in 1930, died in December 1930. The business reorganized under his wife (Hody Hurd) and Johnny J. Jones Jr. and continued operating until 1951, returning to the Florida Strawberry Festival a number of times.
2. Charlotte Rosenberg graduated from Plant City High School in 1930, shortly after becoming the first Strawberry Festival Queen. She attained a two-year degree from Florida State College for Women in Tallahassee and returned to Plant City where she taught third grade at Turkey Creek School. She returned to Florida State College for Women, met and married Charles Rosenberg of Tallahassee in 1935, and received her BS degree in science in 1937. She lived in Tallahassee, working with the family business, Rose Publishing Company, the rest of her life.
3. Misses Irvin Wilder, Elizabeth Carey, and Catherine Fletcher were chosen queen for subsequent festivals. Elizabeth Hull became Miss Florida and was a Miss America runner-up in 1937.
4. James W. Henderson was one of the charter members of the Kiwanis Club in 1921 and one of the charter members of the Florida Strawberry Festival in 1930. He opened a Coco-Cola bottling operation in Plant City in about 1914; his brother, Thomas N. Henderson, became president of the Coca-Cola franchise in Tampa. Thomas Henderson was the president of the Tampa City Council and once served for seven days as mayor, following the resignation of the sitting mayor. Known as "Coca-Cola Jim," James W. Henderson served as general manager of the festival from 1931 through 1941.

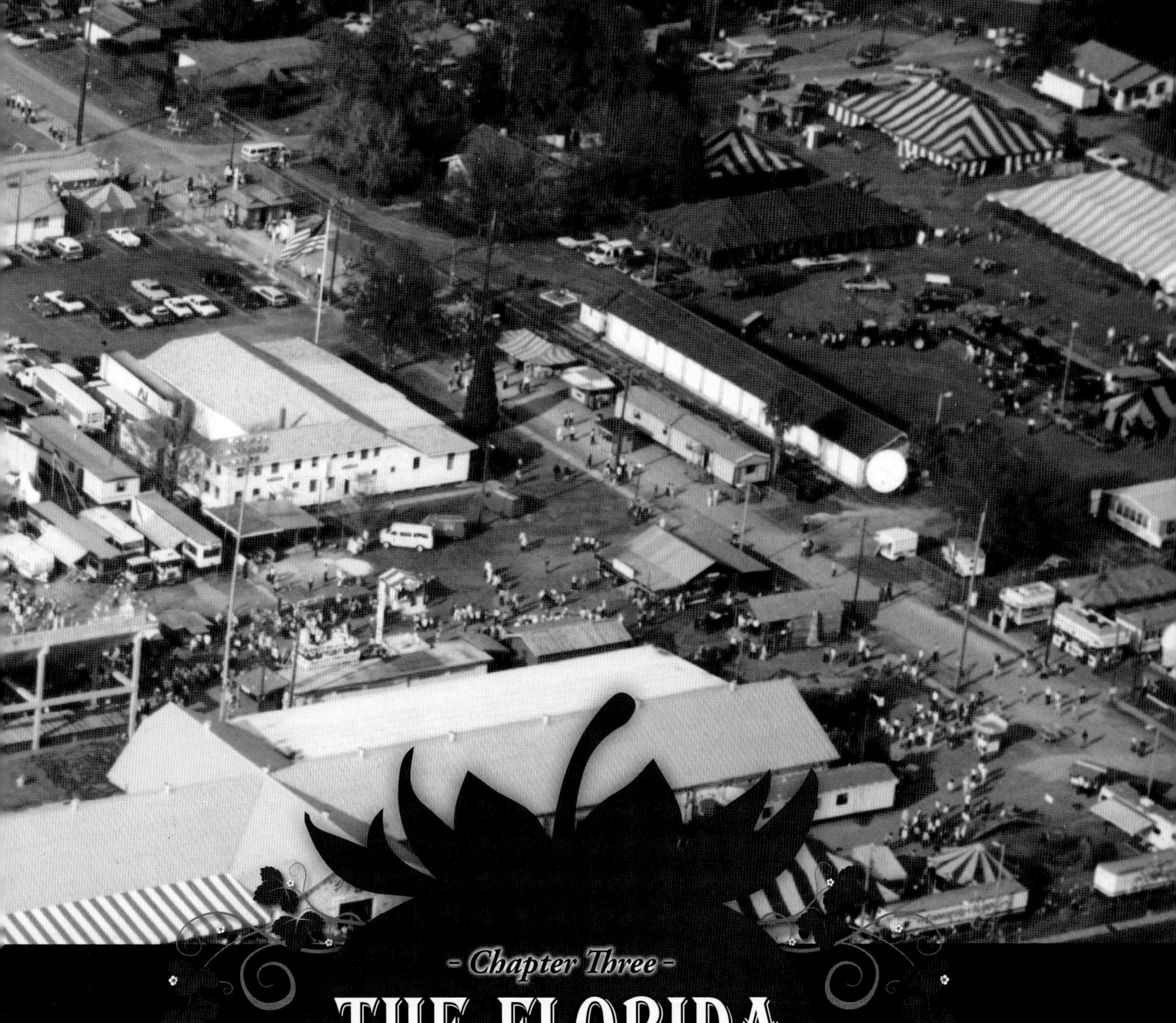

- Chapter Three -

THE FLORIDA STRAWBERRY FESTIVAL

Modern Times

By Lauren McNair

THE 1970S: THE FIFTH DECADE OF FESTIVALS

The 1970s was a period of great transformation for the Florida Strawberry Festival when the annual event became much more than perhaps organizers even imagined. It was a time when family entertainment and amusement took on a whole new meaning. Walt Disney World opened just an hour's drive down the road; *Jaws*, *Saturday Night Live*, and *Star Wars* were popular; and punk and disco music were on top. The festival needed to prove itself both current and traditional and maintain the values for which Plant City was famous. The attendance, entertainment, budget, contests, special events, and more would see quite a boost. It was during this decade that the people of Plant City and festival personnel began to realize what the future could hold for this quaint, local event. Louise Gibbs, who became the staff secretary in 1962 and eventually the first female manager in 1972, perhaps summed up the possibilities best with a memory of her first days in the 60s. She recalled to the *Tampa Tribune* in a 1979 article, "I knew I was sitting on something."

1970

The first festival of the decade, held March 2 through March 7, began with the selection of queen finalists as the primary event of the day. Tuesday, which was Tampa Electric Day, American Legion Day, and Children's Day, was also the day of the glorious Grand Parade. The parade continued to be one of the most talked about and well-attended events during the festival, drawing the likes of state politicians, royalty, churches, civic groups, bands, twirlers, and more. The *Courier* described it this way: "Strawberryland's most spectacular 'confection' – the Festival parade – 'iced' with glittering floats and pretty girls, will sparkle through Plant City's streets Tuesday to the tune of toe-tingling music by high-stepping bands." The parade included over one-hundred floats; a record-breaking twenty-two marching bands; and appearances by Governor Claude Kirk, Florida Senate President Jack Mathews, Florida House Speaker Fred Schultz, State Comptroller Fred Dickinson, Insurance Commissioner Broward Williams, and Commissioner of Agriculture Doyle Conner.

Wednesday was Merchant's Day and Civic Club Day. At the annual civic club luncheon at Johnson's Restaurant, Dr. Harold "Hal" Brewer was named the 1970 Citizen of the Year. It appears

An advertisement was placed in the February 26, 1970, edition of the *Plant City Shopper*.

this was the last year in which the luncheon would be included as part of the festival's events. That evening featured a fashion show by BeeLine Fashions, and Thursday's main event was the coronation of a new queen. Krysta Nifong was selected the 1970 queen along with first maid Karen Campbell and court members Brenda Maggard, Maruchi Azorin, Tobe Robinson, Yvonne Chaney, and Debra Williams.

Visitors saw an influx of youngsters running around in frilly dresses and costumes on Friday with the Baby Parade bringing in what seemed like every child in town under the age of five. The horseshoe pitching tournament and a gospel songfest would follow later that day.

The final day, all Hillsborough County and Tourist Day, saw the conclusion of the horseshoe pitching tournament. All week, visitors were invited to enjoy rides on the Blue Grass Shows' midway and the numerous exhibits available for viewing and participation.

It was in these days that educational exhibits took great priority with the following presenting information or demonstrations: A.C.E Junior Museum Pioneer Booth, Turkey Creek High School, Plant City Senior High School, Pinecrest Future Homemaker's Exhibit, Ruskin County Agent Horticulture Bee Exhibit, County Council Homemaker's Extension Service, 4-H Club Exhibit, Florida Forestry Service, Florida Conservation Commission, Hillsborough County Nutritional Service, County Association for Mentally Retarded Children, and Brandon Art Center. Agricultural exhibits included the Florida Federation of FFA, Strawberry Marketing Commission, Parkesdale Farms, the Lithia, Knights, Bealsville Community Club, Turkey Creek Booster Club Strawberries, and FFA chapters from Turkey Creek, Pinecrest Tomlin, and Plant City High. Commercial exhibits included Seaboard Coastline Railroad, Fletcher Motors, Sharpton Chevrolet, Revels Ford Tractor Company, McGinnes Lumber Company, Tampa Music Company, Frantz Oil Cleaner, Plant City Auto Supply Company, First Federal Savings & Loan Association, C & S Paint Company, Niagara Therapy Manufacturing Company, BeeLine Fashions Inc., Citgo, Do Do Sales & Hobby of Seffner, World Books, Gatlyn Sisters Trio Music Making, Cameo Products, Bay Ford Tractor Company, and M-B Equipment Inc. Other exhibits included BPW Hobby Show, Plant City Garden Club Show, East Hillsborough Democratic Club, Plant City WPLA Radio, Plant City Chamber of Commerce, Kiwanis Club, and the City of Plant City.

Citizens from nearly every business, organization, and corner of the community pitched in to ensure an enjoyable time for families and an accurate reflection of the town's unity and pride.

The 1970 official brochure promotes the thirty-fifth Florida Strawberry Festival and eighth Hillsborough County Fair.

1971

The following January, the festival received recognition from the state's top government official. With Representative Jim Redman (also a festival director) and Commissioner of Agriculture Doyle Conner at his side, the governor signed the following proclamation on January 14, 1971: "Now, therefore, I, Reubin Askew, by virtue of the authority vested in me as Governor of the State of Florida, do hereby proclaim the week of March 1-7, 1971, as 'STRAWBERRY WEEK' in Florida, and urge the citizens of our State to assist in promoting strawberries and the Florida Strawberry Festival, to the best of their ability."

As the festival formally gained statewide recognition, organizers maintained their focus of southern hospitality when welcoming out-of-town patrons. On the Sunday just prior to the festival's opening, a new tradition began with the "arrest" of a Mr. and Mrs. Tourist. As the story goes, a law enforcement official was enlisted to search for a couple driving a vehicle with an out-of-state license plate. When a vehicle was chosen, the officer would pull the car over as if the driver had committed a traffic violation. Once out of the car, the couple would then be "arrested" and crowned Mr. and Mrs. Tourist—and it was a big deal. The royal couple would be whisked away to a local hotel with a free stay, provided gifts from Plant City merchants, and asked to help open the festival the

The winner of the 1971 strawberry recipe contest proudly accepts her trophy.

Admiral Chester Bender, commandant of the US Coast Guard, rides in the Grand Parade.

next day and ride in the Grand Parade. It was quite an honor.

This year's celebration began on March 1 and ran through March 6. Also being honored this year was Admiral Chester R. Bender, who grew up in Plant City and was the commandant of the US Coast Guard at the time. Later in the week, Sherrie Chambers was crowned 1971 queen with first maid Donna Carpenter and court members Christy Cowart, Peggy Gardner, Barbara Cole, Cheryl Keene, and Roseanna Massaro.

Friday kicked off the annual horseshoe pitching tournament and one of the festival's first livestock shows. The Junior Chamber of Commerce (Jaycees) of Plant City started a youth horse show featuring Quarter horses and Appaloosas. The two groups competed in youth western pleasure, reining, western horsemanship, and English pleasure events. Both the horseshoe tournament and the youth horse show continued through the following day and would close out the thirty-sixth event.

The Appaloosa champion receives her ribbon in the festival's first horse show.

As organizers prepared for the 1972 festival, running February 28 through March 4, they spanned public relations efforts throughout the state and into Georgia. With the help of the Hillsborough Marketing Division, strawberries were provided for out-of-state visitors at three welcome stations throughout Florida and one station near Valdosta, Georgia. Queen Sherrie Chambers even traveled to Valdosta in a promotional appearance for the strawberry giveaway.

Just one week before the gates opened, a strawberry picking contest took place at a local farm pitting local and state officials against one another for a little friendly competition and festival publicity. The tradition would continue for years to come. The precise year the contest first began is unclear, though it is believed to have been around 1966.

Though they didn't know it at the time, 1972 would forever change the course of entertainment at the Florida Strawberry Festival. The vice president of the Plant City Lions Club and chairman of the queen's pageant wanted to bring in a Christian entertainer to be part of the upcoming pageant. He somehow landed Roy Rogers's phone number and called. Much to his surprise, Rogers himself answered the phone. While he wasn't available, he mentioned that his wife, "Queen of the West" Dale Evans, may be interested in the event. Evans agreed to come for a fee of $2,000 plus airfare and expenses for transportation, food, and a hotel room. At the time, $2,000 was a fairly hefty price. He took the offer to his fellow Lions, and he along with nine other members agreed to each put up $200 each (if the whole idea fell flat and they didn't recoup their funds). Chairs were rented from Taylor Rental, and the club charged $5 for these special seats set up in front of the stage.

Evans flew to Plant City and was retrieved by two carloads of Lions and Lion wives. On the way back to Plant City, Evans began complaining of a scratchy throat and wanted a Penicillin shot. The Lions worked it out to take her to the office of Dr. Frank Chambers and Dr. Jack Peacock where she was promptly treated.

Florida Strawberry Festival
QUEEN CORONATION
with
DALE EVANS
"Queen of the West"

Thursday, March 2, 1972
Wm. Schneider Memorial Stadium
Plant City
General Admission Grandstand Only
Adults $1.50
Children $1.00

This 1972 flyer advertises Dale Evans's appearance at the Florida Strawberry Festival Queen Coronation.

The group took Evans to Branch Ranch for dinner that evening and held a press conference the following morning at Hillsboro State Bank. She put on a short show that evening prior to the pageant and was on hand to crown the new queen, Linda Scanlon, with first maid Carole White and court members Jenni Barwick, Crystal Beckum, Cathy Stowers, and Ann Wynn. The night's emcees, Myrle Henry and Gerald Hooker (owner of Hooker's Department Store) wore red coats—the start of the tradition of festival directors wearing red coats. The Lions Club continued to bring entertainers the following two years, after which the festival took over the organization of headline entertainment.

Organizers planned for eleven paid events and entertainers this year that would cost just over $15,000 total. Among the lineup was a new puppet show and gospel groups the Speer Family, the Tribunes, and the Prophets. Between the eleven events they expected to profit about $10,000.

To kick off the festival, Mr. and Mrs. Tourist, the new royalty in town, performed the opening ribbon cutting. A new event, the Diaper Derby, made its debut this same day. Tiny competitors crawled their way to the finish to the sound of their cheering and begging parents in hopes of being named fastest crawler at the festival.

Older children had the opportunity to compete in two contests during the week that would also make their debuts along this time—the mutt contest and Kids Greasy Pig Show. In the mutt contest, youngsters were invited to compete with their pup to win categories such as best costumed dog, best groomed dog, smallest dog, largest dog, and best trick

Turkey Creek High School students monitor the school's "Nature Appreciation" booth.

dog. In the Kids Greasy Pig Show, children bowled one another over in the chase of a slippery, greased pig. Whoever could hold onto the slippery swine long enough would likely emerge dirty, greasy, smelly, and, most of all, victorious.

Another new youth contest was established this year that would arguably become the most competitive livestock show held at the festival. The youth steer show began in 1972 in a most unlikely, last minute fashion. A group of exhibitors was rejected from the steer show at the Florida State Fair because their animals did not meet the minimum cattle grading standard required to show. Festival organizers quickly gathered and decided to form an exhibition at the festival so these exhibitors would be able to show their project and recoup at least some of the funds invested. Seventeen exhibitors showed in the first youth steer show, and Randy Langford, a student at Plant City High School, won grand champion.

Once the dust from the festival had settled, Manager E. O. Davenport announced his retirement and was followed by Louise Gibbs, the staff secretary, who became the festival's first female manager.

Grand champion of the first youth steer show, Randy Langford accepts his winnings from Director Harry Carlton.

1973

The festival's board of directors had been under the leadership of E. N. Dickinson for the last quarter of a century. However, Dickinson relinquished his role as president this year and was succeeded by Jack Dempsey.

Up to this point, the festival had no formal event to signal the beginning of the season's festivities. That changed in 1973 with the creation of the Strawberry Ball, a formal gala where guests socialized, dined, and danced to the tunes of Buddy Morrow and Orchestra. The first ball was held on February 10, 1973, and honored Florida Commissioner of Agriculture Doyle Conner. Eighteen individuals from throughout the community served on the committee, taking care of refreshments, decorations, and more, all to usher in the town's favorite time of year. The ball would become one of the most talked about events of Plant City with upwards of six hundred in attendance some years.

This invitation to the festival's first Strawberry Ball states it was held in the Arthur Boring Building.

The 1973 festival, themed "Love America" and running March 5 through March 10, sought "to revive a sense of patriotism and an appreciation for our great country," according to the premium book. Likewise, the schedule of events featured daily patriotic ideas: Monday was billed "Helmsman of America – One Nation Under God," Tuesday was "Heroes of America – Freedom Fighters," Wednesday was "Heritage of America – Our Past," Thursday was "Heart of America – Agriculture," Friday was "Hope of America – Youth," and Saturday was "Happiness of America – Recreation."

The board of directors set the budget for the event at $43,300. Included in the schedule was the first judged horticulture show sponsored by the Plant City Garden Club, a gopher race, the annual strawberry recipe contest, the horseshoe pitching tournament, the Baby Parade, the steer show, square dancing, rides, and more. A feature exhibit was provided by the Florida Department of Agriculture and Consumer Services—five large Florida insects carved from wood. It was to be the "premier showing of the insects among Florida Fairs," according to the *Plant City Courier*.

In keeping with the new initiative of inviting well-known names, three celebrities were asked to be part of the event's festivities. The winners of the established television show *The Dating Game* served as special guests that week, even cutting the ribbon during opening ceremonies on Monday. Pat Boone, world-renown singer and actor, served as master of ceremonies for a queen's selection that would turn out to be a historical one. Boone crowned Phyllis Head the 1973 Florida Strawberry Queen with court members Peggy Welch, Debbie Pollock, Melanie Frohlich, Cindy Schipfer, Linda Adkinson, Sandra Lee Gollahon, and Cecille Dixon, the festival's first African American court member.

Mothers and their children enjoy the festival's midway, circa 1970s.

Later this same year, the festival sought out applications from community groups to be a shortcake vendor at the next annual event. Manager Louise Gibbs told the *Courier* that the Order of the Eastern Star had operated the only shortcake booth ever housed at the festival, but they were in search for another to accommodate the growing number of visitors. The announcement in the paper read: "Any non-profit, community-minded organizations may submit a written bid for a booth at the '74 Festival . . . The Strawberry Festival Board will consider the following criteria as an integral part of the bids submitted by the respective organizations: The shortcake served shall be made with fresh strawberries only. The whipped cream shall be of the highest quality available and the shortcake shall be of such quantity and quality as to be complimentary to the finished product of the strawberry shortcake." Other qualities including sanitation, service, and housekeeping were also to be considered. It would be later when the lucky organization chosen would be announced.

1974

The 1974 festival, held March 4 through March 9, billed itself as a celebration of neighborly relationships. With a theme of "Happiness Is . . . Lovin' Your Neighbor," each day of the event honored a special "neighbor." Monday was "Salute to Canada," Tuesday was "Salute to Georgia," Wednesday was "Salute to Polk County," Thursday was "Salute to Pasco County," Friday was "Hillsborough County Day," and Saturday was "Plant City Day."

The Grand Parade continued to be the highlight of the festival, still attracting royalty, bands, politicians, community groups, and more. In light of the Grand Parade's success, a Children's Parade, later known as the Youth Parade, debuted in 1974 with Dr. Pall Bearer from television station WTOG serving as grand marshal. The *Courier* said, "The event is designed to allow even the youngest to participate in a festival parade with a route much shorter than the Grand Parade. The route will be from the Farmers Market north to Reynolds and then west to the fair grounds." Entries in the parade included six junior high school bands; over one-thousand twirlers in six baton groups; Native Americans performing a rain dance; Boy Scouts and Girls Scouts; and local football, baseball, and basketball teams.

It was also in these years that the Grand Parade sparked an idea for a parade of smaller proportions. Elementary school students in the area were invited to participate in a mini-float building contest where shoeboxes were decorated to look like miniature parade floats. At the time, all entries were displayed at First Federal Savings & Loan in Plant City, the sponsor of the contest along with the Plant City Woman's Club.

The official program book of the 1974 festival shows what the festival is all about.

The title of festival queen was given to Denise Watts on Thursday night as she was crowned by singer and former Miss Oklahoma Anita Bryant. First maid was Mary Sloan, and court members were Lessie Werner, Lisa Sapp, and Terry

Archbell. Joining Bryant on the lineup of headline entertainers was Bobby Gimby, Jeannie C. Riley (who also served as Grand Parade marshal), the Blackwood Brothers with the Tribunes, Ken Curtis (*Gunsmoke*'s "Festus") and the Frontiersman, and David Houston.

As the attendance increased, so did the demand for strawberries and its sweet counterpart— strawberry shortcake. From the organizations who applied the previous year to assist in the efforts, St. Clement Catholic Church was chosen to be the festival's second strawberry shortcake vendor and began serving their "Make Your Own" shortcakes to visitors at the 1974 festival.

Entertainer Lynn Anderson poses with the queen and her court.

1975

Continuing with themes of tradition and gratitude, the 1975 festival, spanning March 3 through March 8, was designed around the theme "I Remember When . . . " Days were devoted to recalling former decades: Monday was "Turn of the Century," Tuesday was "Roaring 20s," Wednesday was "Dastardly 30s," Thursday was "Frolicking 40s," Friday was "Frivolous 50s," and Saturday was "Sobering 60s."

To kick off the event in a grand and sweet way, Main Street Ice Cream Parlor built the towering "World's Largest Strawberry Sundae" by the main gate and, much to patrons' enjoyment, distributed its contents freely once it was completed. Scheduled to entertain throughout the week were Jerry Clower, Jerry Reed, Rick Nelson and the Stone Canyons, Lynn Anderson, Ken Curtis, and Championship Wrestling with Eddie and Mike Graham.

Drawing from the theme, specialty fashion shows entitled "Yesterday, Today and Forever" were held twice during the week, showcasing the evolution of style throughout the last seventy-five years. Hooker's Department Store also sponsored its own fashion shows during the six days.

The 1975 festival program included a photo of Queen Denise Watts.

STRAWBERRY RECIPE CONTEST

CONTEST CLOSES AT NOON
FEBRUARY 21, 1975

RULES:
This contest is open to anyone, except the grand prize winners in all previous contests and employees of The First National Bank in Plant City and the Tampa Electric Company. All entries become the property of the bank to be used in any manner they desire. Entries may be submitted on official entry form or on plain 8½x11 inch paper, clearly written and bearing the entrant's name, phone number, address, and title of recipe. Mail or bring your entry to The First National Bank in Plant City, P. O. Box 1480, Plant City, Florida 33566.

LOOK AT THESE PRIZES!!

FROST FREE FREEZER – Grand Prize
$100.00 Savings Bonds . . . Next 4 Prizes

"COOK-OFF" AT STRAWBERRY FESTIVAL
On Thursday, March 6, the FIVE Finalist will prepare their recipes at the Arthur Boring Bldg. and the Grand Prize Winner will be selected by prominent judges .

Recipes will be judged on:
Flavor (pleasing, identifiable, natural strawberry flavor) 30%; Quality and Appearance (texture, consistency, color and eye appeal) 25%. Recipe Appeal (ease of preparation, ingredients readily available and cost) 25%; Originality (recipe to reflect individual touches, use of strawberries) 20%
Each recipe must contain at least 1½ cups fresh strawberries.

ENTRY FORM
ATTACH, TYPE OR WRITE RECIPE BELOW:

A frost free freezer is the grand prize for the strawberry recipe contest winner according to the 1975 entry form.

On Thursday evening, Lynn Anderson was given the honor of crowning Sheryl Simmons as the 1975 queen. This year a talent competition was incorporated in the pageant so newly elected queens could compete in the Miss Florida Pageant. Martha Lastinger, Rebecca Pollock, Barbara Fulford, and Cathy Johnson were chosen to be in the court.

The week ended with visitors watching Eddie and Mike Graham and other wrestlers duke it out in the ring.

1976

In 1976 all of America was in an attitude of celebration for the entire year—it was America's two-hundredth birthday. With that in mind, the February 26 through March 6 event, themed "Spirit of '76," centered around patriotism. In a welcome letter printed in the annual premium book, Governor Reubin Askew said, "The fair is a vital part of the neighborhood spirit which has always been important to America. It acts as a cohesive factor in cementing relationships in and among the industries, businesses, schools and churches. Hillsborough Countians can truly be proud of this showcase of talent, productivity and achievements which is also an ideal way to observe our Nation's 200th birthday."

In a move partly due to the nation's bicentennial, the board of directors voted to extend the 1976 festival from six days to nine days. It was a bold move. But with an ever-growing attendance, directors sought to relieve the crowded grounds by adding more opportunities for patrons to visit.

On opening day, the first two-thousand visitors through the gate were treated to a slice of birthday cake in celebration of America's birthday. It was also on this night that a new agricultural show for the area's young people started. Just

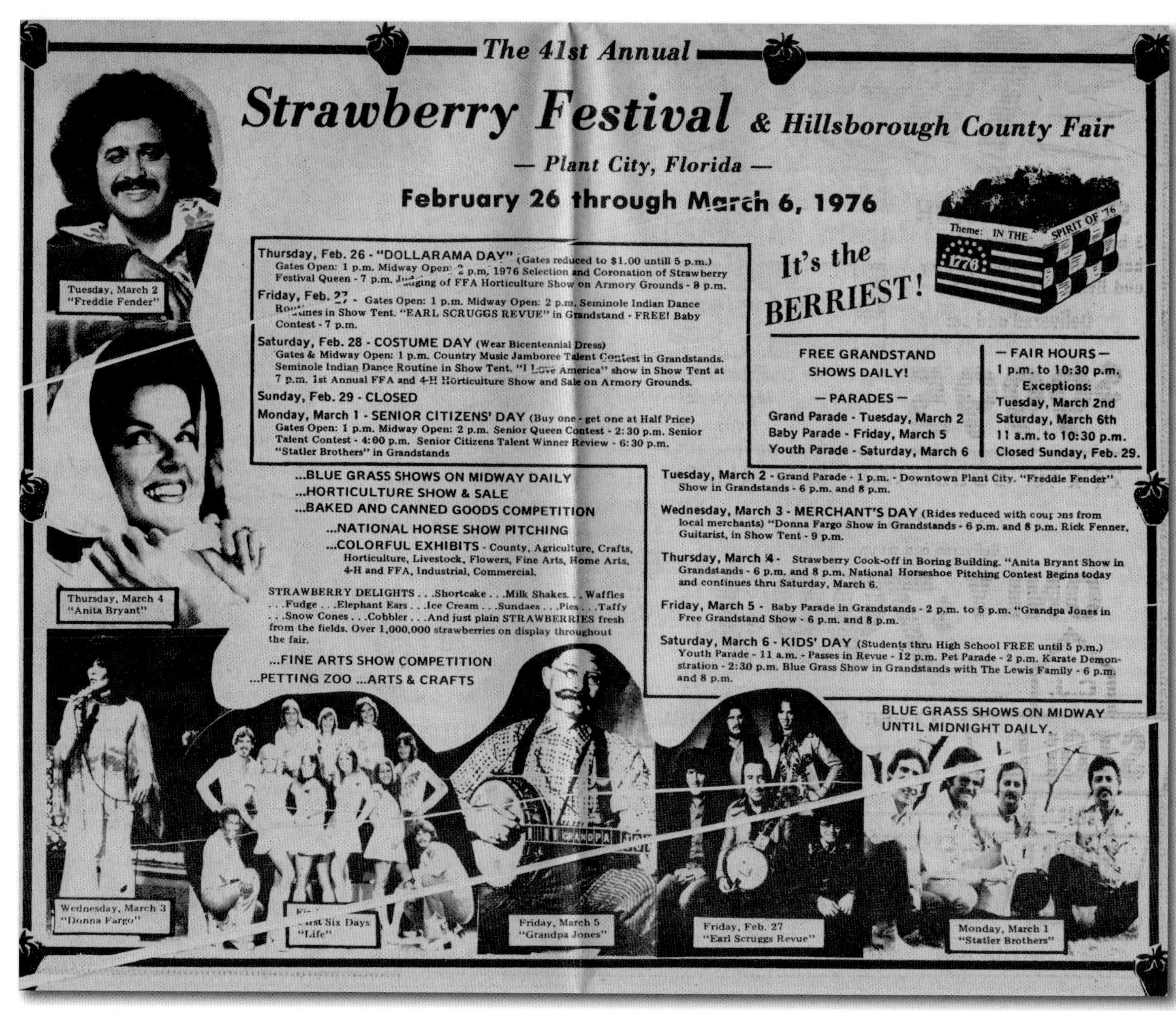

The 41st Annual

Strawberry Festival & Hillsborough County Fair

— Plant City, Florida —

February 26 through March 6, 1976

It's the BERRIEST!

Theme: IN THE SPIRIT OF '76

1776

Thursday, Feb. 26 - "DOLLARAMA DAY" (Gates reduced to $1.00 untill 5 p.m.) Gates Open: 1 p.m. Midway Open: 2 p.m, 1976 Selection and Coronation of Strawberry Festival Queen - 7 p.m. Judging of FFA Horticulture Show on Armory Grounds - 8 p.m.

Friday, Feb. 27 - Gates Open: 1 p.m. Midway Open: 2 p.m. Seminole Indian Dance Routines in Show Tent. "EARL SCRUGGS REVUE" in Grandstand - FREE! Baby Contest - 7 p.m.

Saturday, Feb. 28 - COSTUME DAY (Wear Bicentennial Dress) Gates & Midway Open: 1 p.m. Country Music Jamboree Talent Contest in Grandstands. Seminole Indian Dance Routine in Show Tent. "I Love America" show in Show Tent at 7 p.m. 1st Annual FFA and 4-H Horticulture Show and Sale on Armory Grounds.

Sunday, Feb. 29 - CLOSED

Monday, March 1 - SENIOR CITIZENS' DAY (Buy one - get one at Half Price) Gates Open: 1 p.m. Midway Open: 2 p.m. Senior Queen Contest - 2:30 p.m. Senior Talent Contest - 4:00 p.m. Senior Citizens Talent Winner Review - 6:30 p.m. "Statler Brothers" in Grandstands

FREE GRANDSTAND SHOWS DAILY!

— PARADES —

Grand Parade - Tuesday, March 2
Baby Parade - Friday, March 5
Youth Parade - Saturday, March 6

— FAIR HOURS —
1 p.m. to 10:30 p.m.
Exceptions:
Tuesday, March 2nd
Saturday, March 6th
11 a.m. to 10:30 p.m.
Closed Sunday, Feb. 29.

...BLUE GRASS SHOWS ON MIDWAY DAILY
...HORTICULTURE SHOW & SALE
...BAKED AND CANNED GOODS COMPETITION
...NATIONAL HORSE SHOW PITCHING
...COLORFUL EXHIBITS - County, Agriculture, Crafts, Horticulture, Livestock, Flowers, Fine Arts, Home Arts, 4-H and FFA, Industrial, Commercial.

STRAWBERRY DELIGHTS . . .Shortcake . . .Milk Shakes. . . Waffles . . .Fudge . . .Elephant Ears . . .Ice Cream . . .Sundaes . . .Pies . . .Taffy . . .Snow Cones . . .Cobbler . . .And just plain STRAWBERRIES fresh from the fields. Over 1,000,000 strawberries on display throughout the fair.

...FINE ARTS SHOW COMPETITION
...PETTING ZOO ...ARTS & CRAFTS

Tuesday, March 2 - Grand Parade - 1 p.m. - Downtown Plant City. "Freddie Fender" Show in Grandstands - 6 p.m. and 8 p.m.

Wednesday, March 3 - MERCHANT'S DAY (Rides reduced with coupons from local merchants) "Donna Fargo Show in Grandstands - 6 p.m. and 8 p.m. Rick Fenner, Guitarist, in Show Tent - 9 p.m.

Thursday, March 4 - Strawberry Cook-off in Boring Building. "Anita Bryant Show in Grandstands - 6 p.m. and 8 p.m. National Horseshoe Pitching Contest Begins today and continues thru Saturday, March 6.

Friday, March 5 - Baby Parade in Grandstands - 2 p.m. to 5 p.m. "Grandpa Jones in Free Grandstand Show - 6 p.m. and 8 p.m.

Saturday, March 6 - KIDS' DAY (Students thru High School FREE until 5 p.m.) Youth Parade - 11 a.m. - Passes in Revue - 12 p.m. Pet Parade - 2 p.m. Karate Demonstration - 2:30 p.m. Blue Grass Show in Grandstands with The Lewis Family - 6 p.m. and 8 p.m.

BLUE GRASS SHOWS ON MIDWAY UNTIL MIDNIGHT DAILY.

Tuesday, March 2 "Freddie Fender"

Thursday, March 4 "Anita Bryant"

Wednesday, March 3 "Donna Fargo"

…st Six Days "Life"

Friday, March 5 "Grandpa Jones"

Friday, Feb. 27 "Earl Scruggs Revue"

Monday, March 1 "Statler Brothers"

An advertisement and schedule as it appeared in the February 26, 1976, edition of the *Pasco Shopper*.

four years after starting a youth steer show, a youth horticulture show was founded and open to 4-H and FFA members. Entries in the show were judged this first night and auctioned off in a sale on Sunday evening. The grand finale of opening day was the selection of new royalty, an event previously reserved for Thursday evening. Martha Lastinger was announced the 1976 queen with first maid Karen Leitner and court members Barbara Fulford, Debbie Knight, and Pamela Woods.

Throughout the week, visitors were entertained by the Earl Scruggs Revue, the Statler Brothers, Freddie Fender, Donna Fargo, Anita Bryant, and Grandpa Jones. Patrons were invited to experience the vendors; exhibits of home economics, horticulture, and music; a microwave demonstration; a pet parade with dogs, cats, rabbits, birds, ducks, hamsters, or other pets of choice; quilting demonstrations; and many more entertaining and informational events.

1977

After celebrating the birth of America in 1976, it was only appropriate that the following festival, scheduled February 25 through March 5, be devoted to the way in which the festival got its start—agriculture. Festival organizers explained the reasoning of the theme this way in the premium book: "In an effort to emphasize the importance of Agriculture in Hillsborough County, the 1977 Fair and Festival selected the theme 'Green and Growing . . .' It is the purpose of the Fair Association to continually seek to upgrade the Agricultural Program in the County and State, and to this end we dedicate the 1977 Fair Season."

The 1977 official festival program book features the 1976 queen, Martha Lastinger.

The event did continue to grow, and in 1977 the board set aside one morning of the festival before the main gates opened to welcome the future innovators of the agriculture industry—children. Kindergarteners from schools all over the area got to see and pet barnyard animals, taste strawberries, and learn about agriculture in the first "Sneak Preview."

The festival also continued to grow in its number of royalty. It was in this year that the Junior Royalty Pageant began, a competition for boys and girls between the ages of seven and fifteen. In the inaugural pageant, Robert Julian Clark III and Terri Brantner were selected king and queen; James R. Thompson and Lule O'Neal were selected prince and princess; and David Wayne Kocher and Donna Michelle Hodges were selected duke and duchess. The competition would eventually drop the male division with only a queen, princess, duchess, and baroness being chosen. Competition to bear the queen's crown was still strong. Karen Owens was bestowed the honor when she was named the 1977 queen with court members Barbara Fulford, Amy Carpenter, Julie Willis, and Molly Dull.

The winners of the first junior royalty competition were, from left, Princess Lule O'Neal, Prince James R. Thompson, King Robert Julian Clark III, Queen Terri Brantner, Duke David Wayne Kocher, and Duchess Donna Michelle Hodges.

Around this point in the 70s, strawberry producers began competing against one another at the festival to determine who had grown the finest crop that year. The Hillsborough County Farm Bureau and the festival association partnered to produce the "Strawberry Grower of the Year Competition" with growers competing in the following classes: best flat of berries, best strawberry display, largest berry, most unusual berry, and reddest berry. The winner of the best flat division would be declared "Strawberry Grower of the Year." All entries were auctioned off to benefit the Hillsborough County FFA Foundation, Hillsborough County FFA Federation, James Ranch, Hillsborough County 4-H Foundation, and the Farm Bureau Kidney Fund.

Entertainers were scheduled to perform every night except the night of the queen's coronation. The lineup included Bobby Goldsboro, Mel Tillis, Danny Davis, Bob Harrington, Doug Kershaw, Barbara Mandrell, NWA Championship Wrestling with Dusty Rhodes, and the FSU Flying Circus.

Festival board members and staff stated in the premium book that festivals during this time revolved around the motto of "Potlatch: a South-sea Island word meaning – We give you more than you expect; more than you pay for; and more than you've ever experienced before!"

NWA wrestler Dusty Rhodes appears on the entry form for the fourth annual Youth Parade.

1978

Festival organizers veered this year from a traditional, agricultural, or patriotic theme and instead looked to the future. The premium book stated, "'Challenge of Change' has been adopted for the 1978 Strawberry Festival theme with special emphasis on Energy Conservation. Don't miss the Energy Exhibit being displayed for the first time in Florida Fairs at the Strawberry Festival. Nine State, County and Local Agencies have cooperated to produce an educational exhibit that will portray vividly the 'challenge of change' in the future of energy sources and energy conservation. It is with pride that the Fair Board presents this public awareness program and sincere appreciation is expressed to all those who have cooperated in presenting the 'Challenge of Change' programs." The display mentioned was created through the combined efforts of the US Department of Energy, the Florida State Energy office, the Florida Solar Center, the University of Florida's College of Engineering, Tampa Electric Company, the Plant City High School science department, and the Energy Management Center. It totaled 120 feet and featured information on various energy sources, tips for conserving energy, and insight into the future of energy.

Just prior to opening day, a new fashion show luncheon was established for the ladies of Plant City to preview the latest outfits available at local stores. The event took place on February 21 in the

General admission tickets to the 1978 festival cost each attendee $2.

fellowship hall of First Presbyterian Church and would eventually become a tradition attended by over five-hundred ladies annually.

When the festival opened on Friday, February 24, with a ribbon-cutting ceremony, it was done in a much different location, and elevation for that matter, than ever before. Patrons in the grandstands craned their necks and shielded their eyes to see the opening ribbon cut mid-air by the skydiving group Arch Deal and the Falling Arches. The stunt group landed precisely in front of the grandstands and would continue to appear at several festivals to come. Down on the ground later that evening, Mimi Phillips was selected the 1978 queen with court members Sharon Everidge, Cindy Copeland, Tammy Dukes, and Kim Taylor.

Saturday saw the beginning of a new contest that tested competitors' arms as well as their toughness. The first meadow muffin throw, sometimes referred to as the cow chip throwing contest, was a test to see who could throw a dry pile of cow manure the farthest. These rules were printed in a later edition of the *Plant City Courier*: "Each contestant has two chances to throw within boundary lines. Chips will be at least six inches in diameter. Contestants throwing twice must lick their thumbs for good luck." In the inaugural event, ninety-six competitors dueled for a sought-after gold-plated meadow muffin and the chance to move on to the state champion cow chip toss at the Dade County Youth Fair. As contest coordinators would soon learn, one's background did not necessarily affect the chances of winning. Winners emerged in years to come who knew little about cows and even less about how far a dried pie could fly.

When cow pies weren't soaring through the air, strawberry stems were. Another new event, the strawberry stemming contest, took place later in the week with contestants vigorously working to remove the stems from an allotted number of berries. It was a test more of speed than of skill, and the contest would eventually expand to include a youth category as well.

Visitors enjoyed headline shows throughout the week. Ernie Lee, Jerry Reed, Tammy Wynette, Elvis Wade, the Statler Brothers, Ronnie Milsap, Eddie Rabbitt, and NWA Wrestling featuring Dusty Rhodes all made appearances on the festival stage.

Once the gates had closed on March 4 and all the visitors went home, the board of directors went to work on solving the continuous clash between growing attendance and stagnant land area. Staff of the Hillsborough County Planning Commission suggested the possibility of moving the event to property located south of Reynolds Street, but the idea never came to fruition. Instead, the board voted to construct additional bathrooms and possibly a new exhibit hall to temporarily deal with the cramped grounds.

It was in this same meeting that the board approved a contribution of $5,500 to be used for painting a strawberry on the city's Wilder Road

Tammy Wynette rides in front of the grandstands as the grand marshal of the Grand Parade.

water tower. When city officials voted to not fund the paint job, City Manager Nettie Draughon called on Louise Gibbs. The city had just donated 60,000 salvaged bricks from a construction project to the festival, and Draughon thought it appropriate to ask for something in return. Gibbs told the *Courier*, "The decision wasn't made because of the bricks, but because the board felt like it was something they wanted to do for the city."

1979

In these days, the festival had a relationship with the Clearwater Mall and would often take a day just prior to opening to promote the festival to residents in the area. At the '79 preview, festival representatives handed out cookbooks and strawberry samples and sold discounted tickets. A preliminary celebrity strawberry shortcake eating contest was organized with politicians and media representatives, and Baskin Robbins even hosted a sundae eating contest. The Strawberry Cloggers and the Southern Star Bluegrass Band entertained shoppers, and the queen and court graced the runway in two fashion shows.

Thirteen-year-old Stephanie Silms reacts to her portion in the junior shortcake eating contest.

The theme of the 1979 festival paid homage to the many individuals, talents, and local groups that contributed to the organization of the event. The brochure read:

> With these hands, we in Hillsborough County, have produced the fine array of exhibits, to be seen during the 1979 Fair and Festival . . . It is with pride that we present the 1979 Exposition, and hold forth our hands in a warm welcome to you, the Fair Visitors. We trust you will catch a glimpse of the many Hands united to produce this show, young and old, black and white, male and female. Those of us serving on the staff of this annual production raise our Hands in praise to the God of the Universe who gave us "hands" with which to create!

It was many hands that contributed to the significant expansion of the "Cracker Corner." The Plant City FFA constructed a "Florida Cracker" style pole barn that the J.G. Smith FFA Chapter, Plant City's junior chapter, filled with animals for visitors to see and touch. A historical building from the area was also moved to the festival grounds as a setting for exhibits and demonstrations. The cypress and pine log cabin, believed to have been built in the mid-1800s, was acquired from its original location on Knights Griffin Road. The original structure was constructed using only wooden pegs, and the festival was able to preserve and keep 80 percent of its original timbers. Demonstrations in the building began with Lyndal Toothman showing the craft of spinning cloth from animal hair. As festival organizers worked arduously to keep up with the times, it was also their desire to never lose sight of the area's agricultural past.

But before the festival could even open organizers experienced some headaches and sleepless nights over weather and construction. Louise Gibbs even told the *Courier*, "We're going to open all right, but it may be with lanterns." With original plans to expand the Arthur Boring Building, double the restroom space, and reconstruct the electrical system, it would be one of the most significant expansions the festival had seen yet. But an incorrect measurement, shortage of materials, and late deliveries kept Gibbs "nervous."

The festival did open on time on March 2, just as Gibbs promised. The day kicked off with Paula Fortner, the state champion barrel racer, riding through the crowd with keys in tow to unlock the entrance and signal the grand opening of the festival. Entertainers slated to perform throughout the event were Dottie West (who was later replaced by Jerry Reed), R. W. Blackwood, Marty Robbins, Billy "Crash" Craddock, Jim Ed Brown and Helen Cornelius, Lynn Anderson, the Lewis Family, and NWA Championship Wrestling.

It was in these days that Hillsborough County was home to numerous dairy farms, some saying it was the largest dairy county in the state. With this in mind, Dick Kahelin thought it appropriate to host a youth dairy show that was held for the first time on the second day of the festival. Now that dairy cows were housed on the grounds, they had to be milked. And what better way to accomplish the process than to make a contest of it? The first milking contest took place immediately following the youth dairy show, and it would become highly competitive amongst queens, politicians, and

Circa the late 1970s, a young man gives viewers of the Youth Parade a preview of the upcoming cow chip throwing contest.

others throughout the community. Anyone who wanted to compete, however, was always read the strict set of rules, quoted by Ercelle Smith in a March 1980 edition of the *Ledger*: "The rules of the contest are simple: No kicking, biting or slapping is allowed. That doesn't mean, however, that the animal can't do that to *you*."

It was around this point in the week when a new queen was selected to reign over the forty-fourth annual festival. Pam Smith was crowned 1979 queen in front of the grandstand audience along with court members Lisa Freely, Patti Weyand, Karen Hall, and Teresa Sluder.

The festival had begun beautifully, but the excitement of early activities soon dwindled on Monday when an onslaught of rain threatened the remainder of the week. Some concessions and games closed; hay, sand, water pumps, and drainage pipes simply couldn't keep up with the downpour. Portions of the area's strawberries were damaged, and growers became concerned with whether they could provide the amount of berries needed at the festival. Eventually, the clouds parted. Gibbs was quoted in a 1979 issue of *The American Showman*: "It finally quit on Thursday after four days. You wouldn't believe what we went through. But, the people came anyway and our concessions people were happy." She also bragged on the support from city officials to the *Plant City Observer*: "I shall never forget how City Manager Nettie Draughon came out that miserable night [Monday] to look the situation over. She stood out in the rain, getting drenched from her head down to her boots. That's what I call total commitment."

Once the rain had stopped, the schedule was back on track. The Hillsborough County Cowbelles organized the first beef cookoff contest in which Kurt Weil's beef roulade with mushroom sauce took home the grand prize, and the Turkey Creek Assembly of God continued selling their sweet shortcakes during their first year as a vendor.

The gates of the festival closed for the year on March 10. It was a tough week with emotional highs and lows when complaints would have been understandable. But the end of the decade proved the board of directors, staff, and people of Plant City were resilient and always grateful.

Young viewers of the Grand Parade shield themselves from the week's inclement weather. From left are Jody Hudson, April Shoock, Michelle Heath, Christie Cumbee, Stephanie Tompson, and Sandy Hull.

THE 1980s: THE SIXTH DECADE OF FESTIVALS

The 1980s began to prove the festival was an event known and talked about in upper circles throughout the country. Appearances by political figures, even a first lady, and top-name entertainment meant the event was certainly holding its own. But the 80s also brought with it evolutions in personal entertainment: "Pac-Man," compact discs, the "Game Boy," and the first personal computer were introduced in homes throughout America, and Disney's "EPCOT" opened a mere forty miles down the road. Other options for family fun were being brought right in the doors of homes or in other, grander locations. Could the Florida Strawberry Festival succeed in remaining current while holding to its endearing charm?

1980

In promotion of the 1980 event, neighboring Sea World declared February 22–24 "Strawberry Festival Weekend." While a strawberry festival and a sea lion don't have a whole lot in common, they do attract the same demographic of patrons. To take advantage of the partnership, former queen Karen Owens traveled to Orlando to promote the upcoming festival to Sea World visitors.

The 1980 festival would take place just months before the XXII Olympic Games in Russia, February 29 through March 8 to be exact. Fitting for the occasion, the event was themed "Champions On Parade" with celebrity appearances by Ray Knight (third baseman for the Cincinnati Reds), Minnesota Vikings quarterback John Reaves, players from the Green Bay Packers, Lee Roy Selmon of the Tampa Bay Buccaneers, Bob Mathias (a two-time gold medalist in the decathlon), and other former Olympic champions.

This festival would also find itself to be a timely campaign opportunity for Florida's presidential primary, which was to be held the following week. First Lady Rosalynn Carter visited the festival on Monday, marking the first appearance by a sitting first lady. Mrs. Carter rode in the Grand Parade alongside Florida Governor Bob Graham in a green and white convertible, braving temperatures that dipped down into the twenties and thirties. Republican presidential candidate George H. W. Bush was also scheduled to visit but never made it.

Circa early 1980s, a Florida highway patrolman "arrests" Mr. and Mrs. Tourist.

On opening night, Lisa Harris was selected to reign over the forty-fifth Florida Strawberry Festival with first maid Tina Davis and court members Tami Napier, Holly Smith, and Tamra Coton. Even the rain didn't keep the nearly three-thousand spectators from taking part in the crowning moment.

Throughout the week, visitors enjoyed shows from Jimmy C. Newman and Cajun Country, Dottie Rambo, Danny Davis & the Nashville Brass, Bill Anderson, Larry Gatlin, Championship

First Lady Rosalynn Carter rides in the Grand Parade through downtown Plant City with Florida Governor Bob Graham.

Wrestling, and the Oak Ridge Boys. Duane Allen, a member of the Oak Ridge Boys, recalled a comical story from this first festival appearance in a 1994 article in the *Tampa Tribune*:

> Several years ago, when the Oak Ridge Boys played their first Strawberry Festival, fans presented the group's bus driver with eight cartons of ripe red berries as a token of their esteem. The driver stashed the fruit in the luggage compartment beneath the bus and promptly forgot about it. One afternoon, a couple of days down the road, we went to get on the bus and boy, did it smell ripe. We lost our strawberries on that first run, but now we always look under the bus when we come from the Strawberry Festival because people always fill it up for us.

Clearly, some things did not go exactly as planned. But the week had been a success for other visitors and competitors. The Olympic-themed days concluded with a closing ceremony where the winners of each contest were announced to grandstand spectators and presented a bronze medal for their festival victory. Bob Mathias, director of the US Olympic Training Center and former decathlon winner, spoke about the purpose of the international sports tradition, and patrons

Winners of events at the 1980 festival received this bronze medal.

enjoyed demonstrations of archery, judo, and weightlifting. The festival and ceremony came to an end with a finale of fireworks over the audience.

In June, festival officials and members of the city's planning board met to discuss the festival's expansion plans and need for rezoning. At the time, the festival was zoned residential, multi-family, and commercial—a zoning issue that had simply been ignored by the city up to this point. But rezoning the grounds to community-unit zoning became essential in order to proceed with future construction plans. "In order to make the festival grow properly, we have to zone for what would be conducive to our operations," Gibbs told the *Tampa Tribune*. The newspaper also published exactly what those future plans entailed:

> According to a second detailed site plan submitted to the city for rezoning consideration, future changes included: 1. By 1981, fence the entire portion of the newly acquired property, establish security guard stations, install new turnstiles, and acquire the American Legion Home. 2. By 1982, establish hookup facilities to the carnival parking area, extend a paved track in the midway, build a first-aid facility, move livestock to tents on the north side and make three additions to the Pioneer Village. 3. By 1983, construct permanent animal displays, and install main entry gates on the north and south sides, with turnstiles and office buildings. 4. By 1984, purchase the stadium and football field, cover it and construct a stage. 5. By 1985, purchase the Armory property, build an outdoor theater on the south side, and complete the two-level exhibit building. 6. By 1986, build new restroom facilities and complete Pioneer village.

In this same month, management changes took place. E. O. Davenport became the festival's "acting manager." He would hold the role throughout the next festival.

1981

One of the most well-recognized faces in Plant City was elected president of the board of directors in 1981. Roy Parke, owner of Parkesdale Farms and considered by most a pioneer in Florida's strawberry industry, would serve a two-year term. The board also decided on two-year term limits for board presidents, and they must be re-elected following their first term.

In preparation for opening, the association purchased thirty-two acres of land north of Highway 92 for heavy equipment parking and camping for carnival workers and visitors. An RV park was proposed, but many of the neighbors didn't like the idea and special permission had to be granted from the city, as it did not allow RV parks within city limits.

It was also around these days that the festival would create an event to strengthen their relationship with local media partners. The Media Preview, later known as the Media Party, invited personnel from news stations, radio, newspapers, and magazines to a pre-festival party to socialize, dine, and experience all the festival would be offering that year. In later years, the Media Party would grow to an attendance of over six hundred and include personnel from travel sites, blogs, social media, and more.

Along with a theme of "Showcase of Talent," festival organizers put together a spectacular lineup of the day's popular entertainers. The Bellamy Brothers, Dave & Sugar, Billy "Crash" Craddock, Donna Fargo, the Oak Ridge Boys, NWA Championship Wrestling, T. G. Sheppard, and Sunshine Express all performed to the delight of festival attendees.

Once the festival opened on February 27, judges selected Pam Farris as the 1981 queen with court members Annette Kilgore, Julie Adcock, Lisa Long, and Nancy Wright. The queen and court's attitude, patience, and physical fitness were soon tested when their float broke down during the Grand Parade on Monday and had to be pulled by a truck the remainder of the route. "We're still having a good time with all the other things," Queen Farris told the *Tampa Times*. "This is just one of those things that sometimes happens." "If all else fails," said a member of the court, "we can walk."

The strawberry shortcake eating contest, a long-standing tradition by this time, saw a first in 1981. Kenneth Hickman, a thirty-year-old native of Augusta, Maine, was awarded the famous pig-adorned trophy by consuming the entire shortcake—all three pounds and fourteen ounces— in the allotted time of twelve minutes. Never before had a contestant consumed the entire dessert to claim the victory. Of his win, Hickman quipped to the *Tampa Tribune*, "I knew I could win. Why? Cuz I was hungry." Other contests on the schedule that year were the "Strawberry Harmony" barbershop quartet contest, bubblegum blowing contest, rope jumping contest, rooster crowing contest, Strawberry Rock N' Dance Contest, and many more. They also awarded cash prizes on Senior Citizens Day to the oldest man and woman, longest married couple, best legs, baldest head, and most generations present.

Near the end of the nine-day run, concerns arose over parking and overcrowding issues in addition to the demand for fresh strawberries.

Kenneth Hickman becomes the first competitor to win the strawberry shortcake eating contest by eating the entire dessert.

Both Hinton Farms and Parkesdale Farms had to close their booths early one night for running out of fruit. However, some issues had been at least alleviated by the addition of six-hundred parking spaces, expanded restrooms, and newly constructed entrances at each end of Schneider Stadium Grandstand to help with congestion during concerts.

Things were also a little cramped over in the livestock tent, and organizers of the youth steer show faced great concern over the number of exhibitors. One hundred seventy-seven steers, more than ever before, were entered in the show with fear over not having enough buyers for every animal.

This display shows the entries in the shoebox float contest, circa early 1980s.

After the gates closed on March 7 and the growing pains of the 1981 festival had settled, Plant City found itself to be the subject of a twenty-minute celebratory film starring television personality Peter Marshall. *Florida's Plant City*, a film focused on the city and the role of the festival, was ushered into town in September with a semi-formal gala world premier hosted by the festival association and the Greater Plant City Chamber of Commerce.

In this same month, the board of directors appointed Kenneth Cassels as the new manager of the Florida Strawberry Festival.

1982

The festival has long prided itself on presenting its best: its best personnel, food, entertainment, hospitality, and, of course, strawberries. It was no different in 1982 when "Pick of the Crop" was selected as the theme for the February 26 through March 6 event.

Lined up to entertain festival goers were Doug Kershaw, Paul Lennon, the Muglestones, Charley Pride, Marty Robbins, K.C. and the Sunshine Band, Mickey Gilley with Johnny Lee and the Urban Cowboy Band, Helen Cornelius, and NWA Wrestling. K.C. and the Sunshine Band ended up canceling their performance and were replaced with Dr. Hook & the Medicine Show.

Festival organizers had planned an elaborate schedule with bands, contests, animals, and exhibits by the dozen. And on opening night, a queen was selected to reign over it all. Nancy Wright was crowned the queen and was joined by court members Teresa Adams, Teresa Lott, Patricia Baskin, and Scotti Ray.

The festival made a significant and rare step with the addition of another shortcake vendor. The cherished dessert is well protected, to say the least. But the rarity happened in 1982 when the East Hillsborough Historical Society became the third official vendor of strawberry shortcake "to handle the demand for the dessert" on the festival grounds, joining St. Clement Catholic Church and Turkey Creek Assembly of God, the *Tampa Times* reported. All three vendors sold their desserts for just $1.25 each.

Exactly ten years after the youth steer show had been established, the Livestock Committee decided it was time for another group of animals to compete. In 1982, the youth swine show began, and Peggy Johnson's entry was selected as its first grand champion. In later years, this show would see entries upwards of three hundred, forcing a drawing to narrow the field to around eighty exhibitors.

Around festival time, the hometown event was commemorated in "The Strawberry Song," written by local citizen Mike Mraz. "I just thought it'd be neat to have a song about it," Mraz told the *Tampa Tribune*. "As far as I know, it's the first song ever recorded that's just about the Strawberry Festival in Plant City." Festival Director Al Berry helped Mraz promote the song by playing it on WPLA.

Lyndal "Granny" Toothman spins cloth from animal hair in Pioneer Village.

In June, festival officials announced a reconstruction plan, titled the "1983 plan," to increase seating and restroom facilities in the grandstand area. The plan hoped to boost seating capacity from about 4,500 to about 10,500. It was a valid move as more popular entertainers were booked every year, and more visitors were likely to patronize the shows in years to come.

Competitors in the greased pig contest try to hang on to the slippery swine.

1983

Prior to opening in 1983, the festival association acquired the former American Legion building and completed extensive renovations on the project. It was then decided the Neighborhood Village competition would be relocated from the Arthur Boring Building into the newly renovated facility. "Acquiring the American Legion property has been a major advantage to the festival," Manager Ken Cassels told the *Plant City Shopper*. "It's a place the Neighborhood can more or less call home and will help spread crowds."

When the event opened on February 25, the first five days were not exactly what the board and staff were hoping. The *Courier* reported, "Low temperatures, driving wind and rain made the fairgrounds look like a ghost town over the weekend" and continued about halfway through the Grand Parade on Monday.

Even when the weather is less than desirable, a queen must be crowned. On opening night, Lisa Johnson was crowned queen of the forty-eighth festival along with first maid Suzanne Patterson and court members Kim Deshong, Dee Newsome, and Leslie Brown.

With a growing concern over the public's education of the agricultural process, the festival and the Florida Strawberry Growers Association (FSGA) partnered to bring an educational experience to festival visitors. "Most people think strawberries come from the grocery store. The public needs to know about how strawberries are grown, and this display will help explain that," Festival Manager Ken Cassels told the *Tampa Tribune*. Chip Hinton, then executive director of the FSGA, told the *Plant City Shopper*, "It will tell about strawberry production from the incorporation of a cover crop through bedding the soil, laying the plastic, setting the plants and actually growing the berries. We'll also have an on-the-field grading shack." The group designated a track on the grounds where several rows of strawberries were planted, and festival visitors

An afternoon aerial view of the 1983 festival shows the grandstand and buildings.

The Florida Strawberry Growers Association opens a new exhibit displaying the process of planting, growing, and harvesting strawberries.

could see what the fruit looked like growing on a plant. Farm equipment used in the production of strawberries was also on display with an evolutionary timeline of the differences in planting, growing, and harvesting strawberries then versus many years ago.

Over in the livestock tent, Roy Parke, the longtime festival director and auctioneer of the steer sale, was auctioneering his last one, citing health problems. "I know these kids, they're my kind of kids," he told the *Tampa Tribune* of the children who raise the animals year after year. "They are a special kind of kid, and I've always enjoyed helping them. I hate to quit." Parke had been the sole auctioneer for the sale, which began with the show back in 1972.

The entertainment lineup in 1983 was a big one, stacked with names now known as country's legends: Johnny Cash and June Carter Cash, Loretta Lynn, Conway Twitty, David Frizzell and Shelly West, Louise Mandrell, Bill Anderson, the Muglestones, Paul Lennon, Sylvia, and Championship Wrestling. Many of these same entertainers would return to the festival in years to come.

The festival ended on March 5. Later in the year, B. M. "Mac" Smith Jr. was elected president of the board of directors and would serve for the upcoming two festivals.

Roy Parke takes bids in his final stint as auctioneer of the youth steer show.

Patrons eager to taste the festival's most famous treat visit the East Hillsborough Historical Society's shortcake booth in its second year of concessions.

1984

With attendance numbers in the hundreds of thousands, it was only natural the festival would consider extending the event one more day. They did just that in 1984, the year themed "All Aboard the Strawberry Express," making the Florida Strawberry Festival a ten-day event spanning from March 1 through March 10.

Opening the festival in high-flying fashion was the US Navy's "Chuting" Stars parachuting team who landed near the grandstands and performed other exhibition jumps the same day. Later that evening, Sandra Howard was selected to serve as queen over the 1984 event with first maid Susie Hull and court members Donna Hodges, Roxanne Griffin, and Melissa Pollard.

On Friday night, the midway opened from 11:00 p.m. to 3:00 a.m. in an event deemed "Midnight Madness." This extended riding opportunity was just beginning to take shape and would later be known as "Midnight Magic," "Moonlight Magic," and other names.

This year would see return favorites as well as up-and-comers to the ten-day event's entertainment schedule. Scheduled to perform were the Osmond Brothers, Billy "Crash" Craddock, Razzy Bailey, Ronnie Milsap, Don Williams, Janie Fricke, Ronny Robbins, Conway Twitty, Tanya Tucker, Paul Lennon, the Muglestones, and Championship Wrestling. More patrons would be able to enjoy the shows as well this year due to an additional four-thousand seats being added to the grandstand area.

An elephant makes its way down Reynolds Street during the festival's Grand Parade.

Also with a growing number of visitors came the need for an additional set of information volunteers, deemed "Goodwill Ambassadors." The group, founded and chaired by Festival Director Betty Chambers, included 140 volunteers who were charged with greeting visitors, assisting them with directions around the grounds, and helping with lost and found. "Our main objective is to make them feel welcome, and the second to help," Chambers told the *Plant City Shopper*. "We want each visitor to wish that he had longer to stay or would like a return visit."

The Goodwill Ambassadors, along with other volunteers, would be assisted this year in locating lost items and people by the addition of a designated lost and found area. The Seaboard System Railroad donated a caboose to serve as the headquarters for lost and

A caboose, donated by Seaboard System Railroad, makes its way to the festival grounds to become the lost and found headquarters.

Plant City Dolphins football players and parents pass in front of the grandstands during the eleventh annual Youth Parade.

found items. The piece was delivered via rail to the State Farmers' Market and brought to the east side of the main exhibit hall by trailer. It would be painted bright red to fit in well with the rest of the festival's facilities.

With the numerous additions to the 1984 festival came one unfortunate deduction. With an outbreak of avian influenza in the United States, the commissioner of agriculture cancelled poultry shows across the state. The rabbit show was also cancelled, but this was due to low entry numbers.

Although not in its first year, the festival's "Sneak Preview" event reserved for kindergarten students in the area was a success. Nearly two-thousand kindergarteners from over thirty schools attended a two-hour tour of the festival where they were treated to a petting zoo, sweet snacks, and tours of the festival's exhibits.

Another group of visitors made their way to the festival to tour the grounds and talk with visitors. Much like four years prior, the festival once again found itself to be prime ground for political appearances by Vice President George Bush, Reverend Jesse Jackson, Gary Hart, and Walter Mondale. Vice President Bush enjoyed eating shortcake and watching a clogging performance and spoke to a crowd in the entertainment tent; Reverend Jesse Jackson spoke in the entertainment tent; Walter Mondale toured the livestock tent and spoke in Pioneer Village; and Gary Hart sampled strawberries at Parkesdale Farms' booth.

In line with the effort to ensure the festival remained an educational event as well as an entertaining one were the efforts to include numerous educational programs. By visiting the Educational Activities Center, one could view demonstrations and exhibits on flower arranging, garnishing, dog training, self-defense, pottery, cooking, and more.

1985

In 1985, the Plant City community had two milestones to celebrate: Plant City's one-hundredth birthday and the Florida Strawberry Festival's fiftieth birthday. Fitting for the occasion, the event's golden anniversary was themed "Celebrating 50 Years of Memories" and was held February 28 through March 9.

The momentous occasion also meant the crowning of a fiftieth Florida Strawberry Festival Queen. Kay Newsome was selected by judges to reign over the historic event with first maid Lillian Bowen and court members Michelle Griffin, Patty Harwell, and Donna Hodges.

Denny the Clown entertains guests at the annual media preview.

For the first time, reserved seating tickets were offered for all concerts. Where seating to shows was previously open and offered on a first come, first seated basis, organizers decided to reserve 1,500 of the best seats at every show to be sold for $5 each. About three-thousand tickets were sold on the first day of sales in mid-January. To celebrate the occasion, only the finest entertainers were brought in: Mickey Gilley, Roy Clark, Louise Mandrell, Ronnie Milsap, Tammy Wynette, Loretta Lynn, Lee Greenwood, Larry Gatlin and the Gatlin Brothers, Championship Wrestling, and Atlanta.

The year of celebration was certainly enjoyed by the St. Clement Catholic Church when they reached the 100,000 mark in strawberry shortcake sales. Lani Purcell told the *Tampa Tribune*, "We did 15,000 the first year and thought we were setting the world on fire." The fete was accomplished with the help of 424 church members, 5,123 flats of strawberries, 2,165 gallons of real whipped cream, and 4,950 pounds of granulated sugar.

Loretta Lynn performs for fans during her afternoon show.

Previous winner of the beef cookoff, Marjorie Millsap shows her "Tasty Tenderloin."

At this point, the festival was holding a wide array of contests that allowed visitors of nearly any talent to compete in an event. Festival goers could partake in hog calling, bubble gum blowing, rooster crowing, jump roping, a mutt show, greased pole climbing, senior citizens strawberry picking contest, senior citizens dance contest, and many more.

To help commemorate the year, organizers felt it appropriate to end the festival by honoring a particular group of families whose contributions throughout the past one-hundred years helped build the agricultural industry of Plant City. Twenty families who had owned their central Florida farms and grown crops on them for one-hundred years or more were presented in a ceremony on the final night of the festival prior to the Ricky Skaggs concert.

Once the gates had closed, directors decided against the idea from years ago to move the festival grounds. Instead, they started buying property. The *Tampa Tribune* reported in May 1985 that the association owned the following property in addition to the main office and property on West

Reynolds Street: "Almost all of the two blocks south of Reynolds Street, north of Granfield Avenue, directly across from the Tomlin Junior High School ball park. Most of the block north of Reynolds and west of Ritter Street. Half of the block south of Oak Street, west of Ritter Street. The entire block between Plum and Woodrow Wilson Street, between Lowry Avenue and Granfield Avenue."

In the spring of that year, a historic purchase for the festival and the city was planned when, according to the *Tampa Tribune*, the "Plant City Commission tentatively agreed to sell the Florida Strawberry Festival and Hillsborough County Fair a portion of land where the Florida National Guard Armory" was located. The National Guard planned to move to a new location in the Plant City Industrial Park, and, once moved, the property next to the festival would be owned by the city. The piece totaled nearly five acres and was priced at $208,000.

Around this time, J. Albert Miles took over leadership of the festival's board of directors as president.

1986

The brochure for the 1986 Florida Strawberry Festival, themed "Honoring Our American Heritage," stated the following: "The Fabric of American Life is woven into this Fair and Festival through competitive contests, youth programs, top-named entertainment, parades and a giant ride-filled midway, whatever your interests . . . we have it all."

The festival ran from February 26 through March 8, and the schedule once again included a star-studded lineup. Performing in front of thousands of visitors were Reba McEntire, the Oak Ridge Boys, George Jones, Roy Clark, Waycross Express, Sawyer Brown, Glen Campbell, the Judds, Melody Booth Orchestra, Ricky Skaggs, and Championship Wrestling.

On the first evening, outgoing queen Kay Newsome crowned Rebecca Lewis the 1986 queen. She was joined by court members Stephanie Parker, Kerri Robinson, Pamela Sparkman, and Teresa Tower.

A highlight of the 1986 festival was an animal of a different kind located beside the petting zoo and livestock facilities—Budweiser's famous horses. While the festival has always been a dry facility, there was no harm in viewing their team of beautiful horses. Patrons were invited to view the group of Clydesdales during their several-day stint at the festival.

Robinson's Racing Pigs run for their favorite treat—an Oreo cookie.

Chris Holcomb competes in the Dairy Costume Ball with his cow Marylou, circa mid-1980s.

1987

With each year, the festival's attendance continued to climb. To better meet the demands of this number, the board of directors voted to add another day to the celebration. The February 27 through March 8 event, themed "Pride of the County," ran for eleven days for the first time. Leading the board of directors for the next two years was newly elected president James L. Redman, a former state representative and local attorney.

St. Clement Catholic Church was preparing to sell its one millionth shortcake. In celebration of the sale, the church announced that the winner would win a trip to the Bahamas, a lifetime pass to the festival, and a lifetime supply of strawberry shortcake, among other prizes. Chester Walkowicz from Memphis, Michigan, let a few people ahead of him in line and bought the prize-winning shortcake.

Festival visitors excitedly awaited shows from the Golden Boys of Bandstand (Frankie Avalon, Fabian, and Bobby Rydell), Mel Tillis, Clyde Foley Cummins, George Strait, Cristy Lane, Crystal Gayle (who also served as Grand Parade marshal), Conway Twitty, Reba McEntire, Myron Floren and Orchestra, Roy Clark, John Conlee, Pat Boone, Helen and Billy Scott, and Championship Wrestling.

Chubby Checker poses with clear anticipation of enjoying his St. Clement "Make Your Own" shortcake.

The queen's pageant experienced a significant historical moment with the crowning of the 1987 queen, Rebecca Sue Brown. Rebecca's mother, Ruth Shuman Brown, served as queen in 1954, making them the first mother-daughter duo to both be queen. Chosen to serve alongside Brown were first maid Shayla Wetherington and court members Pamela Sparkman, Stacey Shearer, and Jeanne Roundtree.

A popular contest in these days was the kitchen band contest in the entertainment tent. While the year this contest started is not clear, it was definitely in full swing by 1987. Each group from various communities, many of whom wore matching outfits, was only permitted one regular instrument and all other instruments played had to be homemade.

Reba McEntire meets the 1987 queen, Rebecca Brown.

Tampa Tribune writer Greg Fulton summed up the 1987 festival well with these words:

> After all, where else can you: See 100 cute babies all at once in the baby parade and diaper derby; Hear Crystal Gayle, George Strait, Reba McEntire and many more country stars for $3.50; See cows dressed up and racing pigs; Hear a voice over the loudspeaker crackle twice that keys are left in a locked car, and the engine is running; See politicians throwing the bull in the cow chip toss; See Plant City Clerk Richard Olson emerging from a shortcake line with two long berry stains on his shirt and a dab of whipped cream on his nose.

Terri Parke and Roy Parke present a flat of their own strawberries to George Strait.

1988

As always, the weather is unpredictable, and the start of the 1988 event ended up being less than desirable. The festival was scheduled for March 3 through March 13, and the first few days experienced consistent rain with water trucks eventually having to vacuum water from the property and dirt and wood chips being brought in to cover the ground. The staff and board of directors, led by President M. P. "Bud" Clark, were devastated.

Inclement weather, however, did not stop two things: politicians and royalty. Senator Bob Graham, Representative Mike Bilirakis, presidential candidate Al Gore Jr., and

Couples prove their grace and skill in the senior citizens dance competition.

presidential candidate Michael Dukakis all made appearances throughout the week. Selected to reign over it all as 1988 queen was Robyn Simmons with first maid Dawn Dellapa and court members Christa Moore, Stacey Kendziorski, and Daun Hollins.

Scheduled to perform in front of packed grandstands were Roy Clark, Clyde Foley Cummins, Larry Gatlin and the Gatlin Brothers, Myron Florence and Orchestra (star of the *Lawrence Welk Show*), Exile, Ronnie Milsap, Marie Osmond and Dan Seals, Ricky Skaggs, the Statler Brothers, Charley Pride, Helen Cornelius and the Sweethearts of the Rodeo, and Randy Travis.

At this point, an on-grounds entertainer who would become a long-time festival favorite was first making a name for himself. Dennis Lee, known earlier in his career as Denny the Clown, was establishing his fan base at the festival. Lee first began performing at the festival in the mid-80s with his clown street act, later moving to a gazebo with Denny's Music & Comedy Show and finally to the entertainment tent with his band. Lee would eventually become a name synonymous with the Florida Strawberry Festival and an entertainer whose fans adore him. He has been best known for his outrageous comedy, friendly personality, and love for individuals with special needs. Lee would later say in an issue of the Florida Federation of Fairs' *The Faircracker* newsletter, "It [the strawberry festival] has been magic for me."

As the number of festival visitors grew and along with it the demand for strawberry shortcake, festival organizers and shortcake vendors thought it appropriate to begin offering advance shortcake tickets. While one would pay $2 for a shortcake at the festival, an advance ticket could be purchased for $1.75—and, best of all, it sometimes allowed one to cut in line.

While shortcake customers were getting deals on their desserts, buyers over at the youth steer

Visitors of all ages work to prove their skill in the strawberry stemming contest.

show sale were watching an unbelievable sale unfold when Paul Massaro's steer, Bulldog, was purchased for $22 per pound by the Independent Strawberry Growers of Wimauma. The sale had never seen a number like that, and it would be nearly thirty years before another exhibitor would even come close.

At this point, the headline entertainment program, which began taking shape in the early 70s, was forced to erect a temporary stage before Schneider Stadium for each year's shows. The program was growing and the stage setup needed to follow suit. In the summer of 1988 construction began on a permanent stage for the festival's future entertainers.

Ercelle Smith joins the winners of the annual hog calling contest.

1989

The festival season began much earlier and much differently in 1989 when the sponsors of the queen's pageant decided to hold the selection competition about two weeks prior to the March 2 opening. From this point on, the crowning of a new queen would no longer take place at or during the festival. The 1989 pageant was held across the street in the auditorium of Tomlin Middle School, and Kristen Martinson was selected that evening to reign over the soon-to-be festival with first maid Teresa Tower and court members Jennifer Kubler, Stacee Williams, and Candice Moulton.

It was in this year, with the theme "Agriculture On Parade," that Susan Marschalk, staff writer for the *Tampa Tribune*, recorded a comical comparison of which shortcake vendor had the most delicious recipe for the sweet commodity. Competition between strawberry shortcake vendors has always been friendly, and recipes, presentation, and techniques have always varied slightly. Reverend Joe Woolridge of the Turkey Creek Assembly of God said, "Turkey Creek's is the best." Virginia Hull, former president of the East Hillsborough Historical Society, said, "Ours is the best." And Tilrow Morrison, then supervisor of the St. Clement Catholic Church booth, said the following of their recipe: "You can make your own. It's the best." For most festival visitors, their shortcake loyalty lies with one vendor for life whom they proudly support and recommend to others. No matter one's recipe of choice, visitors

Wynonna and Naomi Judd, grand marshals of the Grand Parade, ride in a white Corsair owned by Ken and Tobe Fleming.

Manager Ken Cassels and Assistant Manager Patsy Brooks present the progress of the festival's permanent soundstage.

devoured about 250,000 of them per year in the 1980s.

When patrons weren't testing different shortcake recipes, they were invited to enjoy a show from Conway Twitty, Indian River, Tanya Tucker, Charlie Daniels, Ray Stevens, the Judds, the Oak Ridge Boys, the Spurrlows, Ricky Van Shelton, Gary Morris, Exile, or Barbara Mandrell. These entertainers were the first to perform on the new stage.

The marshals of the Grand Parade, Wynonna and Naomi Judd, were quite a spectacle as they rode through the streets of Plant City and the festival grounds. They were driven in a white Corsair convertible, owned by Ken and Tobe Fleming, who built a custom white leather seat over the trunk of the car solely for the Judds to sit on during their ride in the parade.

Over in the livestock tent, close to thirty dogs and handlers were competing in the now established 4-H dog show. Animals and handlers were judged on obedience, showmanship, and costumes and competed for trophies and cash prizes.

The festival ended on March 12, and it was during the off-season this year that festival organizers, led by new president William D. Vernon, focused once again on the ever-growing issue of parking by purchasing more land. Ken Cassels said that while the festival didn't push anyone to sell their nearby property, they tried to jump on the parking opportunity when land became available. It was also in these days that directors would begin serious consideration of a permanent soundstage for the festival's continuously growing entertainment program.

THE 1990s: THE SEVENTH DECADE OF FESTIVALS

After proving in the 1980s that the festival was here to stay, organizers were making moves to set the event for decades to come. A permanent stage had already been built, more land would be acquired, and contests and events would be made into traditions. Local citizens were serving in the First Gulf War, and the Internet was taking off. The first sitting US president would make an appearance, and a musical legend would make her last. The investments made during this time would prove themselves beneficial for decades to come.

1990

The festival's 1990 premium book gave a purposeful hint as to what to expect at Plant City's premier event:

"OUR INCREDIBLE WORLD," the theme for the 1990 Festival/Fair can have many meanings. It makes us take stock of all the amazing events, inventions, and discoveries which we see in our daily lives as we enter the last decade of the twentieth century. As we celebrate the 5th annual Strawberry Festival and 28th Hillsborough County Fair, March 1st through 11th, we pay respect to the pioneers of the past and the visionaries of tomorrow. "Our Incredible World" is a fitting theme for a county fair in our area where every day brings word of new developments in agriculture, science, industry, and medicine. The strawberries we grow and enjoy today are products of incredible developments in plant genetics. The way we water and fertilize them also represents progress in the world of agriculture. The chemicals we use are the incredible results of chemical engineering which allow us to grow disease-free, wholesome berries known the world over for their flavor and beauty.

It was in this year that two former queens organized an effort to create a historical walk-through of former strawberry festival queens. They searched through archives and contacted former queens to locate and restore photos from every former queen and court. What ended up taking seven months to accomplish turned out to be a complete compilation of photos from all fifty-four previous queens and courts. The display was originally housed in the Arthur Boring Building but would end up moving around the grounds to create more space for new queens to be added to the display. In 1990, a new set of

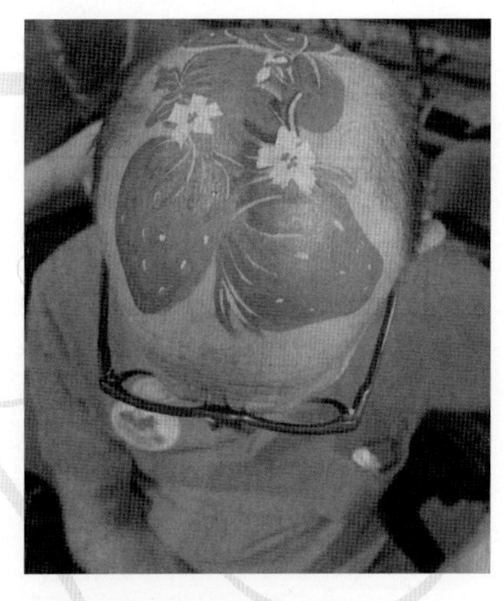

Verl Brant displays his custom strawberry paint job, created by his wife, Jean.

Barbara (Alley) Bowden, 1948 queen, stands with photos while she was queen. Bowden and the 1971 queen, Sherrie (Chambers) Mueller, organized a restoration of the Queens Hall.

names was added to the royal collection—Queen Joanna Cooper, first maid Beverly Forman, and court members Michelle Nail, Lori McGinnes, and Traci Bailey.

Opening ceremonies on March 1 got off with a . . . bang. Hugo Zacchini, known as "the Human Cannonball," brought some daring height to the ceremony when he was shot out of a cannon at a shocking ninety miles per hour. Now that's one way to kick things off!

Because of continuously growing crowds, organizers had to do something about overcrowding on the grounds. To help curb this issue and also provide more space for vendors, they constructed a building just west of the soundstage area capable of holding eleven vendors. The building would become known as the West Grandstand Exhibits. Next door to the newly constructed building, one could enjoy shows from Loretta Lynn, Jerry Clower and Brenda Lee, Ricky Skaggs, Tanya Tucker, Charlie Daniels, Vern Gosdin, Sawyer Brown, Waylon Jennings, Bobby Vinton, Kathy Mattea, and George Strait until closing on March 11.

Rick Gooding, defending champion in the rooster crowing contest, warms up to defend his title.

1991

In a move from historical or agricultural themes to plan the annual festival around, the staff veered from tradition by billing the February 28 through March 10 event "Your Ticket to Fun." Entertainers booked for the fun-filled event were Paul Lennon, the Oak Ridge Boys, Lee Greenwood, Myron Floren, Lorrie Morgan, Ray Stevens, Susan McCann, Charley Pride, Garth Brooks, George Jones, Mickey Gilley, Bobby Vinton, Jimmy Sturr, Jerry Clower and Carl Perkins, and Reba McEntire.

Prior to the festival's opening, Stephanie Chesser was selected queen over the fifty-sixth annual event. First maid was Kim Hursey and court members were Karessa Montgomery, Christy Holt, and Patti Corley.

At the time, the United States was at the end of the Gulf War, and the festival's International

The cover of the 1991 official premium book presented the festival's theme.

Relations Committee devised a plan to show support for those currently serving in the war effort. A booth was constructed in the main exhibit building that featured photos of local individuals currently serving in Desert Storm, and, as festival patrons visited the booth, they were offered ribbons to wear and bumper stickers stating, "Support Our Desert Storm Armed Forces." Visitors were also afforded the opportunity to write letters to those serving, which were mailed daily from the festival's own post office.

By 1991, festival goers were making a tradition of visiting the festival's on-site working post office where their mail pieces could be adorned with a strawberry-themed postage stamp at no extra cost. Some simply wanted the stamp for their own collections.

Later in the week, younger visitors to the festival took part in "Ag in the Classroom" in the livestock area, an event designed to allow children to see, touch, taste, smell, and hear agriculture. Older patrons were invited to take part in "Moonlight Delight" on the midway and ride from 11 p.m. to 3 a.m.

Following the closing of the 1991 festival, board members elected Al Berry, the voice of Plant City on WPLA, as president.

The Wall of Honor in the main exhibit building honored the area's servicemen serving in Desert Storm.

The post office is located in Pioneer Village where visitors mail pieces with an official Florida Strawberry Festival stamp.

1992

Festival organizers seemed to have been preparing for several history-making occurrences at the 1992 event when they decided on the theme "Discover America," a celebration of Christopher Columbus's discovery of America five-hundred years earlier.

In January of this year, the festival purchased the former Johnson's Barbecue building located at the corner of Reynolds Street and State Road 574. President Al Berry said the reason for the purchase was mainly to acquire the fifty-four-space parking lot that went along with the building. This purchase brought the festival's total property to about seventy-three acres.

Just weeks before the festival started, Lisa Diane Stanaland was chosen the 1992 queen. Standing alongside here were first maid Stephanie Goff and court members Trisha Bailey, Monica Brock, and Wendy Maggard.

It would be the 1992 event, held February 27 through March 8, that would mark the first festival visited by a sitting US president. President George H. W. Bush, campaigning for reelection the week prior to the Florida primaries, visited the festival

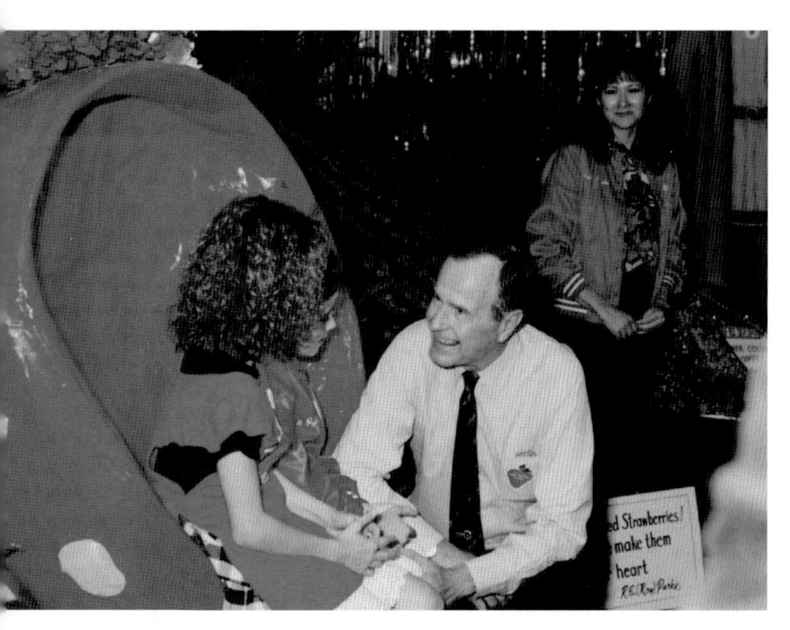

President George H. W. Bush meets two young members of the Parke family at the world-famous strawberry throne in Parkesdale's booth.

for his second time (his first visit was in the 80s as vice president). During his forty-minute stay on Wednesday, he first visited St. Clement Catholic Church's shortcake booth to assemble his custom shortcake, then traveled steps away to Parkesdale Farms' booth to accept a flat of fresh strawberries and strawberry cookies. He finally made his way to the entertainment tent to meet the five finalists in the strawberry recipe contest and talk with contest director Laura York of WFLA Channel 8. The president joked to the *Tampa Tribune* of his visit, "I ate a little shortcake. I could eat it right away. I didn't have to wait for Congress to approve it." While his visit was mostly accepted well, he did give some schoolchildren a tough decision to make. With his visit falling at the same time as one of the pig races, some chose seeing the swine over the president. "I just like pigs. I don't have pigs at home," a Bryan Elementary student said to the *Tampa Tribune*. Others said, "I'll be upset if I can't see the president."

Other well-known individuals scheduled to appear at the 1984 event were Merle Haggard, Charlie Daniels, Shenandoah, Don Williams, Larry Gatlin and the Gatlin Brothers, Barbara Mandrell, Wayne Newton, Vince Gill, Travis Tritt, Emmylou Harris and the Nash Ramblers, and Anne Murray.

Drawing on the theme from the 1980 "Champions On Parade," the festival once again featured an appearance by a well-known Olympian—Mary Lou Retton. Retton, who won the gold medal in the artistic gymnastic women's individual all-around in the 1984 Olympics, met with festival visitors and signed autographs. Patsy Brooks, assistant manager of the festival, said to the *Tampa Tribune*, "It's very exciting. She has tremendous influence for the youth. She is a role model in encouraging them to try their best to achieve excellence. As a sweetheart for youth and adults, she'll bring people in [to the festival]."

During the Grand Parade, Mayor Sadye Martin drew lots of attention as she rode in a red 1935 Ford fire chief's car on loan from Michigan-based Pierce Manufacturing. This would also be the first year in which the Grand Parade was aired on cable television. According to the *Tampa Tribune*, "For the first time, the 90-minute parade, which kicks off at 1 p.m. Monday, will be videotaped and aired on cable by Lakeland television station WTMV, Channel 32." The taping was narrated by Festival President Al Berry, Sandee Sytsma of Parkesdale Farms, and Haywood Henson, host of the weekday country music and talk show *AM With Haywood Henson*.

The strawberry shortcake eating contest saw even more competition than usual when organizers invited representatives from the Plant City Police and Fire Departments to go head-to-head in the annual contest. Lieutenant Carl Rupp, who would later work full-time for the festival, and Deputy Fire Chief Wesley Rounds, who won the competition in 1972 and would later serve as a seasonal employee for the festival, agreed to the match-up. The two poked fun at each other with Rounds telling the *Tampa Tribune*, "I hope he's a better shot with his pistol than he is at eating shortcake." Rupp simply replied, "I'm going to smoke him." In the end, Rounds bested Rupp by eating two pounds and twelve ounces, while Rupp consumed two pounds and seven ounces.

In July of this year, the board of directors voted to drop "Hillsborough County Fair" from the festival's name. Ken Cassels told the *Courier* that "termination of the county fair designation was because the baby parade and queen's scholarship pageant could not be carried out countywide."

Later in this same year, Cassels urged the board of directors to extend the event to twelve days, reported in the *Tampa Tribune* as saying "an extra day could offset poor attendance in cases of bad weather." The board, however, voted down the proposal at the belief that adding another day would be inconsiderate to volunteers.

Deputy Fire Chief Wesley Rounds and Lieutenant Carl Rupp of the Plant City Police Department, pictured in the *Plant City Courier*, prepare for their duel in the strawberry shortcake eating contest.

1993

In early 1993, the festival's board of directors voted to make a sizable donation to ensure the future of the strawberry industry. The University of Florida's Agricultural Research and Education Center was given $25,000 for the purpose of developing a new variety of Florida strawberry. The festival has contributed $20,000 to UF's strawberry research projects every year since this initial donation.

Since its inception really, the Florida Strawberry Festival has provided something different to its visitors—a different feeling that is often difficult to explain or specifically identify. But festival organizers attempted an explanation with the 1993 theme "A Touch of Magic." Patsy Brooks told the *Ledger's Timeout*, ". . . you can touch the magic of the Strawberry Festival with the livestock, you can feel the magic with the excitement during the fair, you can share the magic with your friends out here and you can experience the magic while riding on the midway." The *Plant City Courier* added, "It's as if by magic that seemingly overnight the tents go up, the exhibit halls are decorated, the rides and games are put into place, and the livestock barns are filled."

Contributing to the magic was the introduction of the festival's first mascot—Mr. Berry. The character wore a band leader-type uniform and hat with the focus being on his strawberry head. He was officially introduced at the Greater Plant City Chamber of Commerce's contact breakfast on February 10. While his look has evolved throughout the years, his popularity has skyrocketed. At present, Mr. Berry hosts his own meet and greet three times daily during the festival to shake hands, dance, and pose with visitors.

The festival's ticket office introduced a mail-order system in which patrons could purchase

The festival's official mascot, Mr. Berry, made his official debut on February 10, 1993. He is pictured with festival staff, from left, Ellany Johnson, Ken Cassels, Carolyn Eady, Patsy Brooks, Christy Meyer, and Stephen Tracy.

The 1993 queen and her court with John and Therese Reardon, the fifty-eighth visitors through the gate on the fifty-eighth festival's opening day.

tickets via mail one month prior to the opening of the on-site ticket office. Tickets cost between $4 and $8, and by the time the office opened in early January nearly half of the tickets were already gone. In addition to the expanded ticketing system, the festival broadened the midway and exhibit spaces by purchasing land on the southeast corner of the grounds to Lemon Street. They also installed a new jumbotron screen above the soundstage to give concert goers in the back a better view of the artists to come: the Statler Brothers, Billy Ray Cyrus, Trisha Yearwood, Wynonna Judd, Vince Gill, Diamond Rio, Billy Dean, Charley Pride, Mark Chesnutt, Doug Stone, and Ronnie Milsap.

Just prior to opening day, Ashley Moody was crowned queen of the fifty-eighth annual festival. Chosen alongside her were first maid Julie Henderson and court members Tonya Morrow, Rachel Smith, and Jennifer Smith.

When the gates to the festival opened on February 25, first-time patrons John and Therese Reardon were quickly pulled from the line. Unknown to them, the festival was planning to honor the fifty-eighth guest in line, and John was it. For choosing the best spot in line, the Reardons were gifted with a key to the city, shortcake tickets, cookbooks, coffee mugs, tickets to the Statler Brothers' show, gate admission tickets, tickets to Billy Dean's show, a flat of strawberries, and more. Therese told the *Tampa Tribune*, "This is quite an occasion."

The Grand Parade continued to be one of the festival's feature events with the *Tampa Tribune* saying, "Who needs Fat Tuesday when Sweet Monday is around the corner?" About one-hundred entries traveled the two-mile route with Derrick Gainer, a former Dallas Cowboy and Plant City native, leading the group as grand marshal.

After the gates closed on March 7, the association acquired about three additional acres of land east of Highway 92 and Seminole Lake Boulevard and another half-acre southeast of Reynolds and Violet Streets. This brought the festival's total property to about eighty-four acres. It was also in these days that Joe Newsome was elected by his fellow directors to serve as the board's president.

1994

The people of Plant City and the visitors to come always looked forward to the festival season with great anticipation. Thus it was only appropriate that a festival finally be given the theme "Excitement Is In the Air." Ken Cassels told the *Tampa Tribune*, "We've more than doubled the on-grounds entertainment this year."

Excitement was certainly in the air for five young ladies chosen as ambassadors for the upcoming year. Amy Swilley was selected queen of the 1994 festival with first maid Kellie Heth and court members Cassandra Howard, Brittany Boothe, and Ann Poonkasem.

With a mix of festival favorites and new names, the on-grounds lineup scheduled for the March 3 through March 13 event included Dennis Lee in his own tent, country singers Sandi Powell and Claudia Nygaard, Mike Snider of the *Grand Ole Opry* and *Hee Haw*, the Robinson Family, Susan McCann, Grampa Cratchet, Southern Star Bluegrass, Willie & Company, John Pearson and the Travelling Bell Wagon, Cheyenne Stampede, Kachunga and the Alligator, and the Reppies. Over on the big stage, visitors enjoyed shows from the Oak Ridge Boys, Doug Stone, John Anderson, Wynonna Judd, Wayne Newton, Tanya Tucker, Barbara Mandrell, Ricky Van Shelton, Billy Ray Cyrus, and a *Grand Ole Opry* Legends show featuring Connie Smith, Johnny Russell, Ronnie Prophet, and Little Jimmy Dickens.

By now, the entertainment show tent, later known as the showcase tent, was an established tradition, inviting visitors young and old to be the star for a short show. Robert Griffin, who had been coordinating the tent for twelve years, told the *Tampa Tribune*, "Some of the people who appear there, in my opinion, are just as good as the people who are on the main stage. And some of them should have stayed home and continued singing in the shower." Unfortunately, the strawberry recipe contest, in existence since the 60s, was dissolved this year when no sponsor could be found.

Not far away on the grounds, the sound of squeals was coming from both people and swine as the stars of Robinson's Racing Pigs, one of the most beloved festival traditions, raced to the finish line. Visitors who took in the show were at risk of getting caught up in the comedy of watching Hammy Faye Bacon, Britney Spare Ribs, or one of the others run—and sometimes swim—for the prize of an Oreo cookie. Lucky ones in the crowd were chosen to represent their section, cheering on one pig in hopes of going home with a winning ribbon and a memory. Robinson's Racing Pigs got their start about 1986 and have appeared at most festivals since.

Billy Ray Cyrus sings to fans in his packed afternoon show.

It was in 1994 that the *Courier* ran an article

Visitors watch demonstrations of pioneer skills in Pioneer Village.

highlighting the support and vitality of the festival's two thousand-strong volunteer program. Patsy Brooks said, "They are committed to making sure everyone has a good time at the festival. We could never pay our volunteers for all they do for us." The event's volunteer program would eventually reach about 2,500 individuals of all ages who would welcome visitors through the gate, in the media center, backstage, into the bus parking lot, and so much more.

1995

Nineteen ninety-five would be the event's sixtieth anniversary, with a celebratory theme of "Diamond Jubilee: We'll Dazzle You Daily." In a daily celebration of the March 2 through March 12 occasion, a drawing was held each day for a quarter-carat diamond necklace. To participate, visitors were tasked with visiting twelve specific locations throughout the grounds to have a "passport" stamped. Completed passports were placed in a box from which to draw a winner.

Courtney Lea Clark was crowned the queen along with first maid Angela Williams and court members Delana Hinson, Shelly Nickerson, and Schyler Pickern. In addition to all the duties

The 1995 Florida Strawberry Festival is officially open for business with a ribbon cutting by the 1995 queen and court, Mr. Berry, and members of the board of directors.

Wayne Newton visits with 1995 court member Angela Williams and her pig during the 1995 youth swine sale.

Young visitors enjoy seeing entries in the youth rabbit show.

required of the queen and court, the newly selected young ladies could look forward to meeting Wayne Newton, Pam Tillis, Kathy Mattea, Billy Ray Cyrus, Bobby Vinton, John Anderson, Joe Diffie, Randy Travis, Charlie Daniels, Marty Stuart, and Kenny Rogers.

While the festival had always prided itself on being a family event, there had not been, until this point, a midway area specifically designed for young children. With this in mind, the Strawberry Patch Kiddie Korral (what would later be known as simply the Kiddie Korral) was created. Nick Viscomi, manager of the Mighty Blue Grass Show, told the *Courier*, "We came up with the idea in an effort to offer the young kids a quieter, safer place and their parents some relief from the hustle and bustle of the main midway. The same kinds of rides were available on the main midway in previous years. But we felt there was a need for the smaller kids to have their own area . . ." The Kiddie Korral was constructed on the southeast corner of the grounds and has never changed locations.

To allow more visitors to take part in a headline show, a new set of bleachers was added to

Budweiser's Clydesdales make their way through the festival grounds during the Grand Parade.

the east side of the soundstage with enough seating for 1,500 people. Additional bathrooms located close to the stadium were also constructed. With this addition, the famous "potty passes" allowing visitors to keep their seats when needing a "break" during shows were no longer necessary. Ken Cassels summed it up best when he told the *Courier*, "We haven't reached Utopia around here yet, but we've made definite improvements."

When patrons weren't at headline shows, they were invited to take part in acts from the Fox Brothers, Southern Star Bluegrass Band, Dennis Lee, the Robinson Family, John Pearson and the Traveling Bell Show, Mike Snider, Robinson's Racing Pigs, Grampa Cratchet's Puppet Show, Willie and Company, and Country's Reminisce Hitch.

As was tradition at this point, two years had passed since the election of a new board president, and it was time for the torch to be passed. After the festival in 1995 Terry Ballard took over the leading role on the board of directors.

1996

Nineteen ninety-six was the year the Summer Olympics would be held in Atlanta, Georgia, and festival organizers joined in the excitement of the nation by giving the festival the theme "Reachin' for the Gold: A Gold Medal Celebration." In honor of this, a jewelry giveaway took place each day, much like the diamond giveaway the previous year. This time, winners were awarded gold strawberry pendant necklaces.

The festival would soon run from February 28 through March 10. To kick off the festival season, a new queen was selected to reign over the sixty-first annual festival. Amy Norman was crowned queen with first maid Shea Wooten and court members Emily Dubois, Melonie Wilkerson, and Shelley Causey.

The festival was known at this point for its love of country music. Nearly every headline act fell under the genre of country. Staff members and directors could often be heard saying they knew their audience, and country was what the people wanted. But they decided to shake things up a bit in 1996 and experiment with something new. Organizers booked Roy Clark, Mark Chesnutt, Patty Loveless, Neal McCoy, Mel Tillis, Pam Tillis, Sawyer Brown, Little Texas, Collin Raye, Alabama, and the Temptations.

Contests this year included some of the traditional favorites and some new: the beef cook-off, veggie cook-off, milking contest, cow chip throwing, strawberry shortcake eating contest, strawberry stemming contest, shoebox float contest, and baby contests. The radio station 97 Country also sponsored a clogging contest. It was

Runa Pacha "Indian World" perform South American Indian music for festival visitors.

also in this year that the festival's management experienced a major change. Ken Cassels, manager for thirteen years, announced his retirement in March and was succeeded by Assistant Manager Patsy Brooks. Brooks had been assistant manager for fourteen years.

Ray Clark, Plant City High School agriculture teacher, serves as the long-time emcee of the festival's swine and steer shows and sales.

1997

The festival, still very much an agricultural event, chose "Cultivating and Preserving Our Heritage" for the 1997 theme. Using artwork adorned with gardening tools, old-fashioned ice cream makers, quilts, and plows, visitors were taken back to another time.

Just before the festival opened on February 27, Stephanie St. Martin was crowned the Florida Strawberry Festival Queen. Joining her in festival activities were first maid Heather McDonald and court members DeAnna Blount, Patricia Moody, and Natasha Horn.

As with several former years, patrons over in the entertainment tent were leaving with more money in their pocket than they came with after snagging as much cash as they could in the "Catch the Cash" machine. When visitors weren't fattening their wallets, they enjoyed shows throughout the grounds from the Fox Brothers, Susan McCann, Mike Snider, Dale Jones, Connie Smith, and the Four Lads with the Four Aces. Glen Campbell, Neal McCoy, Billy Dean, Wynonna, Gladys Knight, Barbara Mandrell, Billy Ray Cyrus, Mel Tillis, Aaron Tippin, Lee Greenwood, Jimmy

A festival visitor grabs flying bills during "Catch the Cash."

Politicians, business owners, festival officials, and local citizens eat during the annual parade luncheon at the armory building.

The grand champion winner of the 1997 youth horticulture show.

Sturr and Orchestra with Henry Cuesta, Tom Netherton and Orchestra, and Kenny Rogers would all perform two shows each throughout the eleven-day event.

The festival ended on March 9, and later that year Terry Ballard's time as president came to a close and J. D. Merrill took over as the next board president. It was around this time the festival entered the world of the Internet, purchasing the domain www.flstrawberryfestival.com to host its website. As the popularity of the Internet grew over the years, the Florida Strawberry Festival's site would reach a point where it would be visited over 1.5 million times annually.

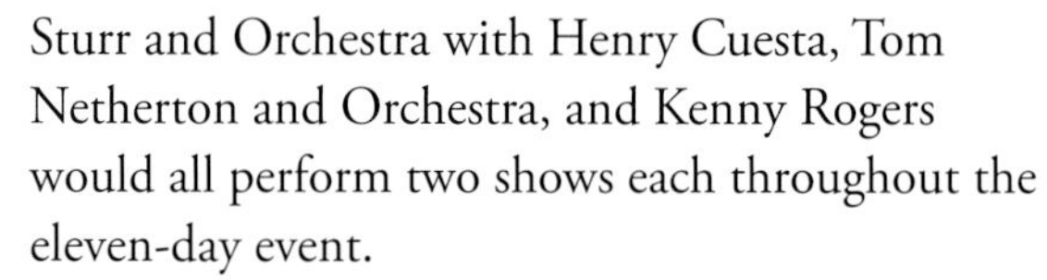

As the festival continued to bolster its headline entertainment lineups and on-grounds entertainment, organizers chose "Catch the Rhythm of the Country" for the 1998 theme. Jessica McDonald was selected the sixty-third queen of the festival along with first maid Amber Farmer and court members Kathleen Guy, Kathie Heth, and Emilie Dubois.

The festival ran February 26 through March 8 and opened on Senior Citizens Day with music and polka dancing that morning by Myron Floren and Jimmy Sturr. The next day, things got a little saucier with the new Sonny's Barbecue charity rib eating contest in the entertainment tent.

The entertainment lineup included, as usual, some of the hottest names in the country. And while the shows and the entire festival ran smoothly, organizers and visitors would realize

Parents line up with their little ones to compete in the baby contest.

Dave McKay and Hank Shaw participate in the inaugural Sonny's Barbecue charity rib eating contest.

Harold Mott and others prepare the meal for the annual volunteer appreciation dinner.

about a month later that this lineup had been much different than ever before. Included in the schedule were Neil Sedaka, Sammy Kershaw, Blackhawk, John Michael Montgomery, the Oak Ridge Boys, Lou Rawls, Neal McCoy, Loretta Lynn, Tracy Byrd, Louise Mandrell, and Randy Travis. In an unfortunate turn of events, Loretta Lynn became ill and could not perform as scheduled on Thursday, March 5. She was replaced by Tammy Wynette, and, unbeknownst to anyone at the time, this would be the final live show for the First Lady of Country Music. Wynette died just one month later when she suffered a blood clot to her lungs. Manager Patsy Brooks told *Amusement Business* this of Wynette's appearance at the festival: "She came here without a signed contract, which is unusual in the industry. She did a good show and talked a lot about her family, which the fans appreciated. Now they realize they were a part of country music history."

Tammy Wynette performs in her final live show before her death one month later.

1999

The Florida Strawberry Festival has always prided itself on providing "southern hospitality" to all of its visitors. To highlight this welcoming environment and the current era, organizers chose "Turning the Century, Southern Style" as the theme for the February 25 through March 7 event. The premium book said, "It's Southern Style . . . And that means entertainment, contests and good food with all the elegance, charm and flair of the South. When it's Southern, it's good, down-home fun for the entire family."

Loretta Lynn, Trace Adkins, Marty Stuart, Beach Boys Family and Friends, Billy Ray Cyrus, Ben Vereen, Neal McCoy, Wayne Newton, Bryan White, Terri Clark, and Vince Gill were all invited to join festival patrons in the southern celebration. Kayla Drawdy was chosen the 1999 queen along with first maid Kristen Parke and court members Elizabeth Raburn, Charleene Closshey, and Jami Waters.

Amidst meeting the entertainers and visitors, Drawdy and her court, along with directors and staff, met chef Emeril Lagasse when he traveled to the festival for a special strawberry-themed cooking segment. Likely unknown to Emeril, he was surrounded by lamb shows, acts of ventriloquism, hog calling contests, pioneer demonstrations, and much more.

Once the festival was closed for the year, the board of directors began planning for the next century by first electing Ray "Rolly" Rollyson Jr. to be the festival's new president.

Chef Emeril Lagasse films a television segment featuring Plant City's famous strawber

A young visitor to the petting zoo reacts to meeting one of its residents.

Justin Gill responds to winning top honors in the youth beef breed show.

THE 2000s: THE EIGHTH DECADE OF FESTIVALS

Around the turn of the century, the city of Plant City was evolving. More chain businesses made their way to town, and new schools meant more families. It was an exciting time for the city and festival and presented opportunities for the event to expand its audience. The festival would begin an era of construction and technological strides, and the queen and court program would gain national attention more than once. Darker moments in the country would also change security at the Florida Strawberry Festival forever. It was a decade to set the festival on a fresh trajectory while keeping its traditions alive.

2000

As the festival was ushered into a new century, the staff, board, volunteers, and visitors had much for which to be grateful. They were also focused, however, on the possibilities of the century to come. It was decided to balance both past and present with this invitation in the premium book:

> "Capture the Memories" for 11 spectacular days at the Florida Strawberry Festival 2000 with some of the best family fun you can imagine! Take snapshots of family fun as you explore a wonderland of Florida crafts, horticulture, livestock, agriculture, photography and art exhibits, parades and special events! Picture the fun as you whirl through the sky on one of our 60 thrilling rides . . . The musical scene is energy-packed with the world's best lineup of top artists . . . You'll have delicious memories of the mountains of home-grown Florida strawberries, served up Festival-style, plump and juicy, red and ripe, and bursting with flavor.

A lifetime memory was created weeks before the gates opened for Alison Archbell. She was crowned queen of the sixty-fifth annual event alongside first maid Erin Welch and court members Kristen Conte, Amanda Hall, and Jennifer St. Martin.

Once the gates opened on Thursday morning, March 2, and visitors rushed the grounds, Jack Harris started his annual tradition of hosting *Harris & Co. Live* from the entertainment tent. The tent would be the set for his show multiple days of the festival. Harris also continued the tradition of "Catch the Cash" during his broadcast. During the first weekend, visitors were treated to a clogging contest, diaper derby, decorated diaper contest, baby features contest, milking show and dairy costume ball, and the youth plant sale.

A highlight throughout the event in the entertainment tent was a stunt group called the "Chinese Acrobats." Audience members held their

The Chinese Acrobats perform acts of balance during their daily show.

breath as they watched the team perform feats of balance and jump to soaring heights. The acrobats would return to entertain festival visitors for several years to come.

Throughout the week, the Statler Brothers, Kenny Chesney, Lee Ann Womack, Kenny Rogers, Terri Clark, Ronnie Milsap, Jo Dee Messina, Mel Tillis, Pam Tillis, Ricky Skaggs, and Martina McBride all graced the soundstage to perform to crowds of thousands. During the remainder of the week, guests enjoyed the lamb jumping contest, a decorated wheelchair contest featuring residents of a neighboring senior home, petting zoos, the beef breed show, and more. Festival organizers turned the lights off on March 12, ending another successful event.

Youngsters practice their showing skills in the lamb show's pee wee showmanship.

Kenny Chesney performs in an afternoon show.

2001

In the early 2000s, the festival was excelling at balance. Attendance, entertainment, and opportunities for youth involvement abounded. All the while, festival organizers were able to maintain the event's famous hometown charm amidst all the growth.

Excitement for the upcoming festival kicked off with the announcement of the 2001 theme "Past to Present," honoring the event's remarkable evolution since 1930. As the March 1 through March 11 festival drew near, it was also time to select new royalty. Kellie Hinson was selected 2001 queen with first maid Lindsay Roberts and court members Ashley McDonald, Deanna Clemons, and Evie Simmons.

At this point, events for youth could be found all over the grounds. Everything from dairy and lamb costume balls to art shows invited the festival's youngest patrons to join in the fun. In the dairy, beef, and lamb shows, the pee wee showmanship contest was a hit. Kids barely old enough to walk took charge of cows and lambs in their best effort to show like their big brothers or big sisters. With their parents' help, of course, youngsters got their first taste of being in the show ring vying to win a blue ribbon.

Visitors took in shows all week from Alabama, Lee Greenwood, Loretta Lynn, the Oak Ridge Boys, Randy Travis, Charley Pride, Aaron Tippin, George Jones, Roy Clark, Neal McCoy, and Ricky Van Shelton. But it was Neal McCoy, known as the festival's favorite showman, who attempted to seal a spot as festival royalty. Much to the delight of those at his show, McCoy removed his own cowboy hat to borrow Queen Kellie Hinson's crown and sash to don during part of his performance.

For several years, children had been enjoying the laughs and fun of the Buck Trout Puppet Show, an educational experience for kids about the great outdoors. When they weren't at Buck Trout's show, they could make their way to the livestock area of the grounds for a AgVenture Farm Tour with a variety of barnyard animals to see, feed, and pet including cattle and sheep.

Competitors in the 4-H Dog Show prepare to enter the ring for competition.

Neal McCoy dons Queen Kellie Hinson's crown and sash during his show.

Kids enjoy the rides in the Strawberry Patch Kiddie Korral.

It was all about family—family fun, education, and values. This driving force and reputation would continue to follow the festival for years to come.

In a board meeting later that year, Robert Trinkle was elected president of the festival's board and would lead the group for the next two years.

2002

The 2002 Florida Strawberry Festival was scheduled for February 28 through March 10. Early in the festival season, Shannon Davis was selected as the 2002 queen with first maid Amber Kosinsky and court members Toinette Gerena, Holly Stein, and Amanda Adams. Davis and her court were the first group under the direction of then associate director Sandee Sytsma. Sytsma had long been encouraging the board to build the program by allocating more funds and scheduling more appearances. The board agreed, and Sytsma became the coordinator of the queen and court program in 2002.

As the festival opened, visitors anticipated many shows. Mel Tillis, the Charlie Daniels Band, Trace Adkins, Toby Keith, the Oak Ridge Boys, Lee Ann Womack, Wayne Newton, Glen Campbell, Brad Paisley, Tracy Lawrence, and Vince Gill all appeared to the delight of festival goers.

The 4-H Dog Show had taken off at this point, with dogs of every breed, size, and color showing off their skills and beauty in obedience, breed characteristics, and more. Watching a pack of canines bark and compete directly next to the steers was quite a sight. Throughout the week, the Baby Parade, strawberry shortcake eating contest, and petting zoos, among others, continued to be some of the festival's most beloved and anticipated events.

Just six months after the September 11 terrorist attacks, festival organizers were forced

This afternoon view shows the crowd on the festival grounds.

to take a hard look at their security plans. A major change was to alter the route of the Grand Parade; no longer would it go through the grounds of the festival but instead skirt the outer perimeter. Security at the annual event would be forever changed as a result of September 11.

Barbara Caccamisi, director of the St. Clement "Make Your Own" shortcake, with the winner of the 2002 strawberry shortcake eating contest.

2003

During this time, the country was arguably more united and patriotic than ever. Things were no different in Plant City, as festival organizers wanted the annual event to reflect this sense of pride with the theme "Let Freedom Ring." Chosen to reign over the February 27 through March 9 event was Queen Erica Der with first maid Brandie Johnson and court members Allison Bethea, Shana Norris, and Kaley Mercer.

The *Palm Beach Post* published a feature story during the festival that summed up the personality of the event quite perfectly. The story reads:

> Even before you pass through the gates of the Florida Strawberry Festival, you get the sense this isn't your typical fair. It's a feeling of hospitality that comes from the middle-schooler who guides you to your parking spot and sends you away with a hearty, "Have fun!" Or the security staffer who practically apologizes for having you submit to the requisite bag search. But once you enter, the down-home flavor really registers . . . Mac Smith said, "We don't have a beautiful town with a lot of lakes. We hang our hat on the strawberry." But the key, say insiders, is that it retains that down-home sensibility.

A little girl enjoys being royalty for a moment in Parkesdale Farm's royal throne.

The 2003 festival did not lack big name entertainment. Invited to be part of the hometown event were Ronnie Milsap, Terri Clark, Ricky Skaggs and the Del McCoury Band, Vince Gill, Tanya Tucker, Brad Paisley, Randy Travis, Bobby Vinton, Neal McCoy, Billy Ray Cyrus, Martina McBride, Jimmy Sturr and Orchestra, and Myron Floren and Orchestra. Festival guests got a treat of a different kind with on-grounds entertainers Raggs Kids Club Band, ventriloquist Terry Fator, "Runa Pacha" Indian World, and others.

Cheerios partnered with the festival on opening day to promote their new triple-berry cereal blend featuring, of course, strawberries. With a giant spoon in hand to commemorate the collaboration, Cheerios personnel gave out free samples of the new mix during Jimmy Sturr's opening morning performance. Later in the week, a giant Cheerio and a giant strawberry rode together in the festival's Grand Parade.

When all the shortcakes had been eaten and the shows ended, another festival was in the books. Plans for next year began with the election of Kenneth Lucas as the festival's new board president.

Brad Paisley performs during an afternoon show on the soundstage.

John Pearson and the Travelling Bell Wagon perform for festival patrons throughout the grounds.

2004

In 2004, the festival truly was "Florida's 'Berried' Treasure"—a premier family event in Florida and a hidden gem gaining more notoriety by the year.

Prior to its February 24 opening, Kaitlin Sharer was invited to be the face of the celebration when she was crowned queen. Joining her were first maid Ashley Pippin and court members Leanna Blake, Lyndsey Terry, and Crystal Wiggins.

Among the seven livestock shows, seven contests, and countless vendors and exhibits, well-known performers were booked. Patrons

Long-time entertainers at the festival, the Oak Ridge Boys pose backstage before their show.

were entertained daily by Larry Gatlin and the Gatlin Brothers, Trick Pony, Tracy Byrd, Lonestar, Mel and Pam Tillis, George Jones, Kenny Rogers, Bobby Vinton, Brenda Lee, Rascal Flatts, the Bellamy Brothers, and Wynonna Judd.

While the festival had been known for decades for its love and inclusion of livestock and farm animals, another group of creatures had established themselves as a popular exhibit among visitors. Phillips Exotic Petting Zoo featured animals that most would otherwise never see in their lifetime: giraffes, kangaroos, zebras, emus, and more. Located just north of the Hull Armory Building, the animals were housed under, as one would imagine, a pretty tall tent.

The grounds were quiet after the March 7 closing. Later in the year, festival organizers took out their hard hats for the first time in many years. The former American Legion building that had been housing the Neighborhood Village was torn down to make way for something new. Once the site was prepared, construction began on a new exposition building to house vendors, restrooms, and a place for royalty to reside.

The newly constructed expo building is located east of the main soundstage.

2005

Construction on the new building was completed in early 2005, just prior to a dedication ceremony. What stood before the board of directors, staff, and other guests that day was a 15,000-square-foot metal building with thirty-two-foot high ceilings and enough space to house about seventy vendors during the upcoming festival. It was a big step in upgrading the festival's property and would spark more construction projects within the next ten years.

But the vendor space is only what is visible to the public. Hidden deep inside the building is a place where royalty resides known as "The Palace." Each year, the festival's queen, first maid, and court members are provided their own "home away from home" to prepare for each day's events. It's a place for hair, makeup, snacks, and even the occasional nap. Ashley Watkins, selected as queen shortly after the expo building's dedication, was the first queen to enjoy the comforts of "The Palace." Watkins was joined by first maid Caycee Hampton and court members Catie Walker, Brooke Ellis, and Amy Stewart.

It had been seventy years since the gates of the Florida Strawberry Festival were opened for the first time, and so the theme "Seventy Years of Fun and Still Jammin'" was quite appropriate. Visitors to the 2005 event would indeed be "jammin'" to the likes of the Oak Ridge Boys, the Charlie Daniels Band, Aaron Tippin, Louise Mandrell, Clint Black, Vince Gill, Wayne Newton, Neal McCoy, Neil Sedaka, John Michael Montgomery, Michael W. Smith, and LeAnn Rimes. This would be the first time a contemporary Christian artist was included in the lineup, starting a trend that would continue through to the present.

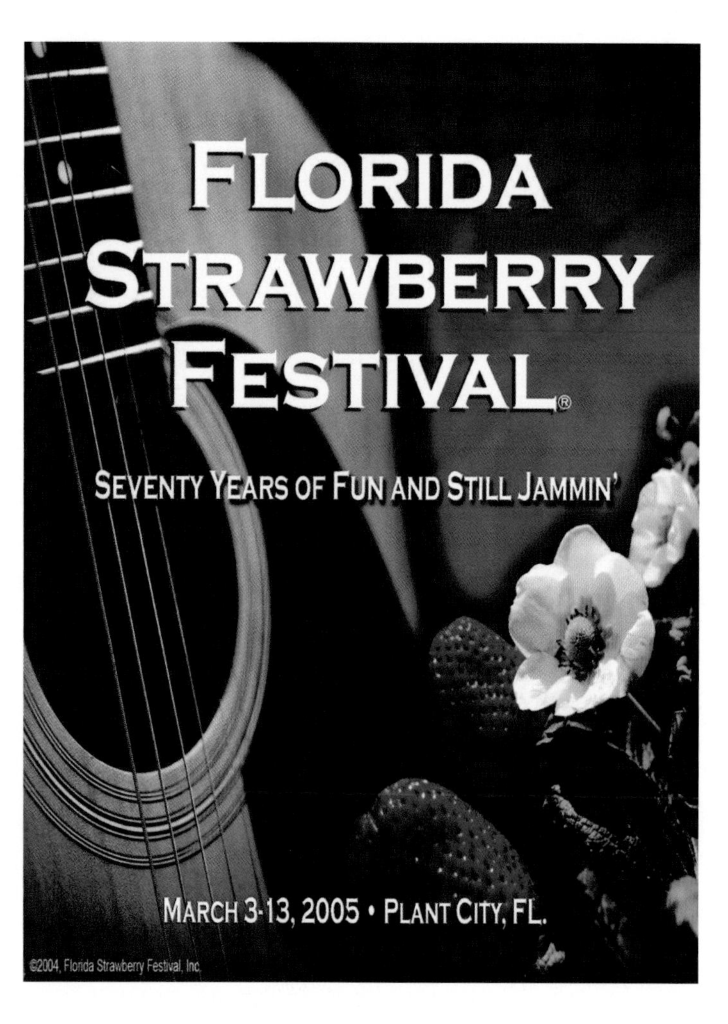

Jimmy Sturr, an eighteen-time Grammy winner, entertained some of the first visitors on March 3 with his signature show beckoning audience members to leave their seats and join in the polka. Sturr had been a long-time staple entertainer on Senior Citizens Day and continues to entertain festival goers to this day.

Following the eating contests, livestock shows, concerts, and fellowship, the 2005 festival ended on March 13, and the board faced its biannual election once again. This time, Johnny Dean Page was elected president of the board of directors.

The official premium book of the 2005 festival joins music with the ever-important strawberry.

2006

In the late 1800s, Henry B. Plant completed a cross-Florida rail transportation system from Sanford to Tampa. Before the tracks would end in Tampa, they ran straight through what is modern-day Plant City, named for Mr. Plant himself. The railroad would serve as a shipping method for the area's strawberries, playing a significant role in Plant City earning the title "Winter Strawberry Capital of the World." In paying homage to the history of the city and its agriculture industry, the 2006 festival was themed "Making Tracks to Fun."

Five young ladies were selected in February to join in the upcoming eleven-day event. Hannah Hodge was chosen the 2006 queen alongside first maid Ilene Chavez and court members Megan Shelley, Kayla Gaschler, and Julie Boback.

The festival was scheduled for March 2 through March 12, and leading the entertainment lineup were Darryl Worley, Terri Clark, Steven Curtis Chapman, Larry Gatlin and the Gatlin Brothers, Mel Tillis, Connie Smith, Willie Nelson, Lonestar, Debbie Reynolds, Dierks Bentley, Jo Dee Messina, Rhonda Vincent, Trace Adkins, and Big & Rich. Although there weren't any train tracks on the grounds, one could certainly see tracks of a different kind. "Boards, Bikes & Blades," a BMX show, made its festival debut in 2006 with an action-packed show full of tricks and turns on skateboards, bicycles, and roller blades. The Rowdy Rooster Puppet Show & Roadster also made his debut, much to the delight of the festival's younger visitors.

Over in the livestock tent, adults were now competing in their own division of dairy showmanship. After watching their children show the day before, parents enjoyed competing against one another to see who retained their showing skills from back in the day. It was friendly, but it was still a competition.

In an unfortunate turn of events, one of the festival's most beloved congregations of animals had to be cut from the program in 2006. Both the barnyard petting zoo and exotic petting zoo were no longer allowed on the festival grounds for reasons of liability. It was a blow to the board of directors and staff, as the festival had always prided itself on bringing visitors face to face with agriculture. But they simply had no choice and were forced to move on. Once again, they were resilient and forged ahead, hoping to make up for the loss.

Interested parties were invited to the twenty-ninth annual Florida Strawberry Festival Fashion Show held on February 2, 2006.

2007

Over the years, much of the talk about the festival revolved around one thing—the combination of strawberries, whipped topping, and cake (or biscuit—depending on preference) known as strawberry shortcake. As the festival continued to expand in its audience and offerings, festival organizers wanted to make it clear that one could enjoy more than just a sweet treat during their visit. With that in mind, the 2007 festival was given the theme "It's More Than Just Shortcake" and was celebrated March 1 through March 11.

As festival organizers continued to plan, they began to run into an entertainment issue that would be both a blessing and a curse.

Shortcake was highlighted on the official brochure of the 2007 festival.

More and more entertainers were not keen on the idea of doing two shows per day, as had been the festival tradition for many years. The Entertainment Committee was faced with the daunting task of finding twice as many entertainers as they were accustomed to booking. While it did provide a greater variety of options for festival guests, it was also a great challenge to now find over twenty headliners for just eleven days. But yet again, the committee was persistent and somehow found twenty-one artists to grace the festival's stage—George Jones, Clint Black, Barbara Fairchild and Bill Anderson, Gretchen Wilson, Mark Lowry, Casting Crowns, Joe Nichols, Martina McBride, Lee Greenwood, LeAnn Rimes,

The newly constructed gate one entrance has sixteen ticket windows.

Ray Price, Randy Travis, Little Richard, Kenny Rogers, the Oak Ridge Boys, John Anderson, Big & Rich with Cowboy Troy, the Bellamy Brothers, Hootie and the Blowfish, Josh Turner, and Montgomery Gentry.

When the gates of the festival were opened on Thursday morning, visitors noticed a much different entry process taking place. Rather than handing their admission ticket to an employee, as had always been the custom, their entry ticket was *scanned*. The festival had begun electronic ticketing. Guests could now purchase admission and concert tickets online, and all tickets on the grounds were tracked electronically. While it sounds like a tremendous technological step, the process got off to a rough start. Networking and signal issues created somewhat of a disaster with the new system. During the next two years, the kinks would be worked through, and the process would prove itself extremely effective. One of the primary benefits was it provided a much more reliable number of ticket sales and attendance to report.

Over in the Music Revue Tent, another group, "Vocal Trash," was using trashcans and other thrown away items to create unique, unbelievable sounds. Festival favorites Dennis Lee, "Runa Pacha" Indian World, the Southern Star Bluegrass Band, Robinson's Paddling Porkers, and Grampa Cratchet were before their loyal audiences.

As festival organizers were faced with booking more acts, it was also a good time to branch out from the norm. Plant City had a growing Hispanic population, and the Entertainment Committee contracted with Groupo Los Nenes, a Spanish group, to perform in the showcase tent. This act was the spark that would start future festival events for the area's Hispanic community.

After the festival, Johnny Dean Page's term as president ended, and it was time for the board to elect a new leader. Gary Boothe was elected board president for the next term of two years.

As the festival had become more technological this year with its ticketing, the board found it imperative that a larger, more aesthetically pleasing facility be constructed for ticket sales. After the festival closed in 2007, they broke ground on a main entrance at gate one with an astounding sixteen ticket windows. The building opened for business in December when tickets for the next lineup of headline entertainers went on sale.

2008

Two thousand eight would prove itself to be a big year for the festival in garnering national recognition. When deciding on "Lights, Camera, Action!" as the theme, it seems as if festival organizers knew that fame was coming.

When Kristen Smith was chosen as the queen with first maid Shaunie Surrency and court members Amanda Sparkman, Jaclyn Raulerson, and Britney Balliet, they gained the typical notoriety of the queen and court: photos and feature stories in local publications and surrounding areas, guest spots on television and radio stations, and photos and appearances at local events. That changed when queen's coordinator Sandee Sytsma received a call from a Pulitzer-winning writer who was interested in writing a feature story on the queen and court to be pitched to several big-name magazines. Anne Hull, a native of Plant City, traveled from New York with a photographer to spend a few days with the girls and record the life of Plant City royalty. Hull's finished product, entitled "The Strawberry Girls," was published in August as a multi-page feature story in the *New Yorker*. The town was abuzz. Never before had the festival or the queen and court program been featured in a publication of this magnitude. This story, as it turns out, would spur more attention for queens and courts to come.

Back on the festival grounds, the gates opened on February 28, and visitors were dancing and singing to the tunes of superstar acts. Jimmy Sturr and Orchestra, the Pied Pipers and Clambake

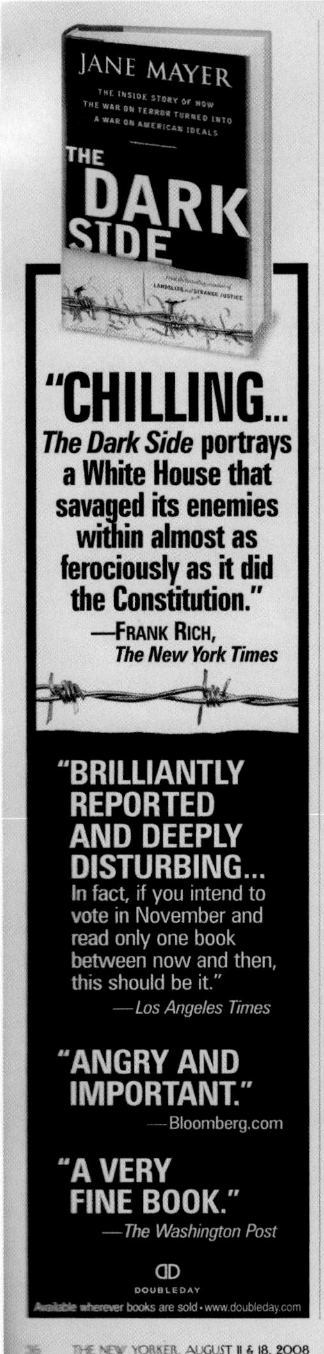

LETTER FROM PLANT CITY, FLORIDA

THE STRAWBERRY GIRLS

Celebrating a new queen and her court.

BY ANNE HULL

One Sunday last February, a young woman named Kristen Smith left the parking lot of Bethany Baptist Church, in Plant City, Florida, and drove along a two-lane country road with a large gold crown on the seat beside her. The mossy pasturelands around Plant City—the winter strawberry capital of the world—were exploding with ripe fruit. Kristen was two weeks into her reign as the 2008 Strawberry Queen, and the crown was already causing severe headaches. It weighed nearly a pound and, even bobby-pinned on top of her thick chestnut hair, left a mark on her forehead, an affliction known among generations of Plant City Strawberry Queens as "the queen's dent." She was on her way to a lunch where she would be making her official début, and she was nervous.

Kristen, who is nineteen, was not a regular on the beauty-pageant circuit. She could eat a plate of ribs and two hours later be craving pork rinds or red-velvet cake. When her spirits flagged, she read Scripture. She drove a pickup truck, attended a Christian college, worked part-time as a waitress, and wanted to spend the rest of her life in Plant City, raising a family. Kristen Smith disproved the theory that the Strawberry Queen had to be the well-connected daughter of a town scion; her father repaired washing machines for a living. She looked like a young Bobbie Gentry, and she was just what Plant City was looking for in these modern times.

Along the road, mom-and-pop operators were selling flats of berries from tents and campers. One of the fruit stands favored by tour buses was doing such a business in shortcake that the whipped cream was pumped out by nitrous tanks. Fifteen per cent of the nation's strawberries are produced in Eastern Hillsborough County between December and April. Still, the Outback Steakhouses and the stucco subdivisions were getting closer, and the life that Plant City celebrated was vanishing, acre by acre. The coronation of a Strawberry Queen had come to seem almost an act of defiance.

Kristen pulled into the Red Rose Inn & Suites, a motor lodge off Interstate 4, whose marquee said "Queen's Luncheon." The purpose of the luncheon was to introduce Kristen, along with four other young women, who had been chosen to serve as members of the 2008 Strawberry Court, to the thirty or so women whose husbands run the Florida Strawberry Festival, in Plant City. These women are called "directors' wives," and, unofficially, they represented the traditions of the town.

The queen and her court made their entrance. Their dresses were modestly cut, their hair was teased, and they wore strawberry pendants, strawberry charms, strawberry bracelets, and strawberry barrettes. With white satin pageant sashes, they stood at the edge of the dining room, more tentative than triumphant. They seemed to grasp the magnitude of their role when one of the luncheon guests clasped her hands and, in a loud voice, said, "Are these *our* girls?"

I grew up in central Florida in the nineteen-sixties, barefoot half the time and running around the orange groves where my father worked. I remember flocks of white birds that would lift from the backs of cattle, disturbed by the jackhammers and bulldozers clearing land for Walt Disney World. Disney would never have what Plant City's Strawberry Festival had; we had the smell of hay and manure, the crêpe-paper float that, on parade day, carried five young women through archways of Spanish moss. I must have been seven or eight when I got to ride in the parade, holding a tinfoil wand and wearing tap

36 THE NEW YORKER, AUGUST 11 & 18, 2008

The beginning of a feature story published in the *New Yorker* about the 2008 queen and court.

The queen and court are seen in the *New Yorker* with a local strawberry grower.

Seven, Bobby Vinton, Josh Turner, Charley Pride, Miranda Lambert, Jeff and Sherri Easter with the Isaacs, MercyMe, Chris Cagle, Alan Jackson, Mickey Gilley, Travis Tritt, Glen Campbell, Blake Shelton, Crystal Gayle, Tom Jones, the Smothers Brothers, the Charlie Daniels Band, Gene Watson, Trace Adkins, Billy Ray Cyrus, Neal McCoy, Sugarland, and Jason Aldean all graced the stage at the 2008 Florida Strawberry Festival.

Across the grounds in the showcase tent, Reverend Art Jones, Minister V. Michael McKay, and the Bible Based Fellowship Cathedral Choir were in the middle of their set. In a show for the faith-based community, particularly the African American community, the group had been leading festival visitors in worship and scripture for several years.

When the festival closed on March 9 and work began on planning the next event, the Lions Club was pondering a big change in the annual queen's pageant. For years, they and festival organizers had received complaints about the swimsuit portion of the competition, hearing that many young ladies refused to compete simply because they did not want to wear a swimsuit in front of their entire town. With this in mind, the decision was made to cut the swimsuit competition in favor of casual wear. The change was mostly welcomed, and the new phase of competition would take effect in the upcoming 2009 pageant.

It was also around this time that Patsy Brooks, manager since 1996, announced her retirement from the festival. Brooks was followed by Paul Davis, a retired major in the Hillsborough County Sheriff's Office who had worked in security at the festival for nearly thirty years.

2009

As organizers were preparing for the festivities of the upcoming festival, the theme "Hometown Salute to America" was selected for planning the eleven-day event. When the season got underway, the Plant City Lions Club held the annual queen's pageant, not knowing history was about to take place when Lauren Der was crowned as the Florida Strawberry Festival Queen. Lauren's sister, Erica, served as queen in 2003, making them the first set of sisters to both be queen. Lauren served alongside first maid Sara Beth Newsome and court members Joclyn Emerson, Megan Cochran, and Morgan Feaster. Not long after their selection, their coordinator received a phone call from an ABC producer in New York who had read the story in the *New Yorker* and was interested in the festival's royal tradition. After learning more about the group and meeting with the girls, reporter Juju Chang traveled to Plant City with the producer and a crew to film each of the five young ladies in their everyday lives and performing their duties during the festival. Not long after the festival, a five-minute segment aired on *Good Morning America Weekend* introducing the Florida Strawberry Festival and its queen and court to America.

Guests at the February 26 through March 8 festival were treated to shows from George Jones, Larry Gatlin and the Gatlin Brothers, the Nitty Gritty Dirt Band, Kellie Pickler, Jake Owen, Third Day, Rodney Atkins, Taylor Swift, Helen Cornelius with Jim Ed Brown, Randy Travis, Connie Smith with Marty Stuart, Travis Tritt, Brenda Lee, Ray Stevens, Mel Tillis, Lorrie Morgan, Ronnie McDowell, Jeff Foxworthy, Mark Lowry, Kool & the Gang, Julianne Hough, and Jessica Simpson. Julianne Hough was later replaced by Trent Tomlinson, Love and Theft, Chad Brock, and Killer Beaz.

As the festival continued to grow and attract more attention, organizers felt it necessary to focus efforts on media relations. With this in mind, they opened the festival's media center, a professional workspace for media partners to use while they covered special events. The center is housed in the former parsonage of the Plant City Church of God, whose property and buildings were purchased by the festival in 2000. Currently, the center includes office spaces, work space, living space, and a sponsor lounge and is staffed by nearly forty volunteers.

The festival continued to host seven livestock shows—swine, dairy, poultry, rabbit, lamb, steer, and beef. The decision was made to relocate the livestock area for the 2009 festival to the northwest corner of the grounds in preparation for a much larger move down the road.

Following the end of the festival, Mike Sparkman was elected president of the board by his fellow directors and led the group for the next two years. At the end of the year, renovations had been completed on the former Plant City Church of God sanctuary, which was turned into the festival's administrative offices. Staff members moved in throughout the month of December. The festival's office manager and vendor manager, Ellany Johnson, likes to brag that her office is where the old baptistery was located.

The turn of the century brought numerous changes, some good and some less than desirable. But the festival continued to prove it could withstand just about any obstacle and began preparing to welcome visitors for the next decade.

A "Hometown Salute To America" was the theme of the 2009 festival as shown on the official brochure.

The 2009 queen and court are filmed for *Good Morning America*.

THE 2010s: THE NINTH DECADE OF FESTIVALS

By this time, the festival was under the direction of a new manager. Plant City's commerce and population continued to grow, and so did the festival. The grounds would see more permanent fixtures as two major construction projects were completed, more land would be acquired for parking, additional special days were added to the calendar, more businesses would seek out partnerships with the community event, and social media would allow nearly 120,000 individuals to follow the festival's daily happenings. The queen's pageant would see three history-making moments within just four years, and the board of directors would work diligently during these successful years to set up the event for the decades that would follow.

2010

One can likely assume that the twenty-six charter members of the Florida Strawberry Festival couldn't be sure the event would ever see a seventy-fifth birthday. But much to the delight of the community and even the state, it did just that in 2010 with the theme "Come Celebrate With Us!"

Natalie Burgin was selected to reign over the momentous occasion as the 2010 queen. Accompanying her in the festivities were first maid TyLynn Eben and court members Dayla Dementry, Ashlyn Robinson, and Rachel Hallman.

To join in the occasion, the Entertainment Committee brought in a host of big names to entertain festival visitors from March 4 through March 14. Jimmy Sturr and Orchestra, the Guy Lombardo Band, Roy Clark, Billy Currington, Bobby Bare, Julianne Hough, Michael W. Smith, Switchfoot with One Republic, Heidi Newfield, Darius Rucker, Mel Tillis, Bill Engvall, Debbie Reynolds, Josh Turner, Bill Medley with Sam Moore, the Oak Ridge Boys, REO Speedwagon, the Smothers Brothers, Tracy Lawrence, Clay Walker with Lee Brice, Patty Loveless, Sara Evans, and Lynyrd Skynyrd all made appearances on the festival stage.

While the Entertainment Committee had been adamant about booking Hispanic entertainers for several years now, a feature event called "Hola! Plant City" was created in 2010 with prominent entertainers in the Hispanic community. Los Sobrinos De La Lluvia, La Nueva Ilusion, and guest host Jose Luis took part in a debut that surpassed all expectations. The showcase tent was so full of fans that people were forced to stand on picnic tables outside the tent to simply catch a glimpse of the stage.

The entertainers in the festival's first Hola! Plant City performed in the entertainment tent.

The 2010 festival saw the addition of three new themed days, each with its own special guests to spotlight. On Tuesday, everyone was seeing red and purple as the ladies of the Red Hat Society took over the festival grounds in the first Red Hat Day. Members of the social organization served as the festival's special guests for the day, receiving discounted admission and taking part in a welcome ceremony in the grandstands.

The following day a theme debuted that many would begin calling their favorite day of the festival. Wednesday, March 10, was the festival's first American Heroes Day when all active, reserve, or retired military veterans, law enforcement, and first responders were admitted free of charge. It was a day for the festival to thank those who made the greatest sacrifices; a day for military, law enforcement, and first responders to reminisce and show their pride in their area of service and for families and fellow servants to enjoy the company of one another. Visitors took part in the Honor American Heroes Ceremony later in the afternoon, standing with great pride as each branch of military and each group of first responders were recognized. The feeling in the stadium and about the grounds that day was special. Everyone from ticket sellers to vendors to entertainers took time to shake hands and say "thank you." Days like American Heroes Day is what the Florida Strawberry Festival is all about.

This veteran of World War II, joined by his wife, is the first veteran through the gate on the festival's first American Heroes Day.

On Saturday, Farm Worker Appreciation Day, the festival invited employees of local strawberry growers to visit the festival free of charge. The work

The ribbon-cutting ceremony marked the opening of the festival's new administrative office.

ethic of these individuals provided the strawberries sold during the festival, and it was imperative their contributions be recognized.

Other new additions to the festival's on-grounds entertainment included Jay Taylor, Jason Young, Galaxy Girl and the Galaxy Globe of Death, and the Paul Bunyan Lumberjack Show.

The 2010 festival was brought to a close with a new fundraising opportunity established by Stingray Chevrolet in Plant City. The owner of Stingray approached Lee Bakst, the festival's assistant manager, about hosting a raffle during the festival—and not just any raffle. Hurley wanted to give away a brand new car to raise funds for Unity in the Community, a Plant City organization. The deal was sealed, and a new 2010 Chevrolet Camaro sat at Stingray's display on the festival grounds, begging visitors to buy a $5 raffle ticket. The vehicle was given away on the last night of the event, and the tradition would continue in festivals to come. Stingray has since raffled off Camaros, Silverados, and even custom Corvettes. To date, the raffle has raised about $800,000.

Later in the summer, Queen Natalie Burgin's photo was published in *Southern Living Magazine* in a special feature on southern queens. It was the first time the festival appeared in the south's most prominent publication.

2011

Sweets, deep-fried anything, and every strawberry food item imaginable have always been part of the festival experience. Chocolate-covered bacon, corn dogs, cinnamon rolls, elephant ears, and hundreds more items have been sought out by visitors every year. In spotlighting all the festival's delicious treats, the 2011 festival was given the theme "Taste the Flavor of Fun."

As the festival, scheduled for March 3 through March 13, and the queen's pageant drew near, no one knew a rare feat was about to be accomplished once more. Victoria Watkins was selected the 2011 Florida Strawberry Festival Queen, the sister of 2005 queen Ashley Watkins. For the second time in just two years, sisters were crowned queen. Also chosen were first maid Victoria Garren and court members Kori Lane, Summer Norris, and Taelor Highland.

When the festival opened to the public, new acts Brady Goss, Pirates of the Colombian Caribbean, and the Magic of Lance Gifford and Company were entertaining guests throughout the grounds. In the entertainment tent, a new trio of sisters, the Gothard Sisters, took the stage with their Celtic-based, folk sound. Outside the tent, youngsters craned their necks to see the nine-foot-tall creature scooting by. Rock-It the Robot debuted in 2011 as well, talking, dancing, and singing with the young and young at heart. Both the Gothard Sisters and Rock-It the Robot would return numerous times in upcoming festivals.

Rock-It the Robot entertains and towers over festival guests at nine feet tall.

At this point, the festival only had one eating contest—the strawberry shortcake eating contest. Granted, it was extremely popular among guests and media personnel alike. But it was time for something new, and a contest involving corn dogs seemed like a logical choice. On Wednesday, the first super dog mega corn dog eating contest took place with competitors challenged to eat corn dogs measuring two feet long. It was a bit messy, but the popularity of the event has withstood every year since.

An unbelievable twenty-six headline entertainers performed for the festival's guests in 2011. Making appearances were Bobby Vinton, Kenny Rogers, John Conlee, .38 Special, Jeremy Camp, Gaither Vocal Band, Trace Adkins, the Judds, Charley Pride, Clint Black, Chubby Checker, Chris Young, Tanya Tucker, Rick Springfield, George Jones, the Doobie Brothers, Ray Price, Josh Thompson with Justin Moore, Allstar Weekend with Jennette McCurdy, Billy Ray Cyrus, Easton Corbin, Lady Antebellum, Jimmy Sturr and Orchestra, and the Guy Lombardo Band.

Changes were taking place at the festival, and visitors welcomed the fresh appeal. Soon after the event ended, Ronald Gainey was elected president of the board of directors and was charged with helping maintain the festival's new zeal.

Competitors in the first mega corn dog eating contest are ready to begin.

In April, a plan that had been taking shape for nearly thirty years finally came to fruition when ground was broken for an agricultural show center that was to be one of the finest in the state. The plans were monumental: a 21,000-square-foot building for the livestock show arena and two open-air pavilions on each side for housing animals, measuring in at 22,000-square-feet each. What was once a dream was now going up block by block.

The ground-breaking ceremony of the 65,000-square-foot agricultural center.

2012

The festival had long been promoting itself as an event to be enjoyed by the entire family, with no alcohol, entertainment and rides for every age and interest, and endless memories to be made with loved ones. With this in mind, staff and directors gave the 2012 festival the theme "Growing Sweet Memories." It was celebrated March 1 through March 11.

A sweet memory was surely made for Chelsea Bowden and her family when she was crowned the 2012 queen. In yet another rarity for the pageant, the new queen was the granddaughter of not one but two former queens—Barbara Alley, her paternal grandmother, was queen in 1948 and Ruby Jean Barker, her maternal grandmother, was queen in 1953. Chelsea was selected to serve with first maid Calli Jo Parker and court members Chelsea Talavera, Olivia Higgins, and Erica Kelley.

The Entertainment Committee combined both yesterday's and today's favorites. The year's lineup included the Oak Ridge Boys, Gretchen Wilson, the Gatlin Brothers, Crystal Gayle, Demi Lovato, Jimmy Sturr and Orchestra, Ronnie McDowell, the JaneDear Girls, Air Supply, Jake Owen, Hank Williams Jr., Felix Cavaliere's Rascals, Josh Turner, Brenda Lee, Vince Gill, Lee Greenwood with Louise Mandrell, the Band Perry, the Charlie Daniels Band, Luke Bryan, Cornell Gunter's Coasters with Bobby Hendrick's Drifters and the Platters, TobyMac, Allstar Weekend, Rodney Atkins, Easton Corbin, and Reba.

In addition, the on-grounds entertainment was packed with new faces and sounds. The Kandu Magic Show, Lloyd Mabrey, and hypnotist Ron Diamond all made their festival debut. Also on the lineup were the Lumberjills—world champions at axe throwing, chainsaw carving, crosscut sawing, and log rolling. Next to the Lumberjills, another new show, the Sea Lion Splash, was making waves

Members of the World Champion Lumberjills show their skills in a crosscut competition.

A feature member of the Sea Lion Splash performs for festival visitors.

The official ribbon cutting of the festival's newly constructed agricultural center was attended by young and old.

with visitors. Featuring the highly intelligent sea lions, visitors watched them flip, jump, and balance.

The crown jewel of it all though was the completed agricultural show center, which hosted every livestock show beginning with the 2012 festival. The complex was officially dedicated on Thursday, March 8, in the presence of former and current FFA and 4-H members, parents, advisors, teachers, and community supporters. A dream had been achieved.

2013

In 2012, the festival adopted a strawberry into its marketing that would be dressed according to the year's theme. After all, the strawberry was the festival's brand. The character who was dressed as a country singer the year prior now donned a beret and mustache and held a paintbrush and color palette, portraying the 2013 theme "Our Masterpiece of Fun."

Before any masterpiece could be created, new royalty came first. Kelsey Fry was selected the 2013 Florida Strawberry Queen with first maid Ericka Lott and court members Maddy Keene, Madison Astin, and Jamee Townsend.

Throughout the event, spanning February 28 through March 10, every guest was almost guaranteed to find something of interest. For those who enjoyed the comedy of good ol' country folk, there was the Sweeney Family Band Country Comedy Revue. For those who were interested in learning about agriculture and the food production process, there was the SouthWest Dairy Farmers Milking Show. For those who listened to a mix of traditional and contemporary country music, there was Kari & Billy. And for the dog lovers, there was JUMP! The Ultimate Dog Show.

Emily Nichols meets the 2013 queen, Kelsey Fry, as she competes in the annual Baby Parade.

For the eaters in the crowd, there was something altogether different. A new eating contest had been added to the schedule of events just two years prior, and festival patrons seemed to enjoy both watching and participating. With that in mind, organizers added two more opportunities to eat one's way to victory. The first fried corn on the cob eating contest, held on Friday, pitted

Competitors find some comedy amidst the first strawberry mashed potato pie eating contest.

competitors against one another to eat one ear of fried corn the quickest. Exactly one week later, participants were challenged with eating one pie plate of mashed potatoes topped with a strawberry, of course, in the strawberry mashed potato pie eating contest.

Across the grounds on the soundstage, patrons could see Chubby Checker, Foreigner, Gene Watson, Casting Crowns, Gloriana, Justin Moore, Randy Houser, Alan Jackson, Lorrie Morgan with Pam Tillis, Dwight Yoakam, Bobby Vinton, Martina McBride, Neal McCoy, Trace Adkins, Mel Tillis, Brantley Gilbert, T. G. Sheppard with Janie Fricke, Scotty McCreery, the Gaither Vocal Band, Jimmy Sturr and Orchestra, Steve Hall and the Shotgun Red Show, Bret Michaels, Hunter Hayes, and Blake Shelton. The lineup was top-name acts, yet a blessing and curse. For the first time, tickets to two headline entertainers were sold out—and scheduled for the same day. The shows for Hunter Hayes and Blake Shelton sold every seat available, with Hayes scheduled to perform at 3:30 p.m. and Shelton at 7:30. With two sold-out shows and it being the final day of the event, 99,000 guests entered the gates that day. Traffic, maneuvering the grounds, and even making it to one of those shows were difficult, to say the least. But the obstacle proved to be a lesson for festival organizers that would help improve traffic flow for years to come.

Taking on the task of leading the board of directors in the upcoming years, James "Jim" Jeffries became the festival's twenty-fourth president. One of the first tasks Jeffries and the board decided to take action on was the construction of another building on the grounds. For decades, the Neighborhood Village and the competition and display of baked goods, preserves, quilts, clothing, and other homemade items was forced to move to several locations throughout the grounds. With none of them being ideal for the display, it was time to design a building specifically suited for this display of tradition. That fall, construction crews began work on the 12,000-square-foot building that would house the Neighborhood Village, the horticulture show, and the Hall of Queens.

Construction of the new Neighborhood Village building is underway.

2014

The 2014 festival, held February 27 through March 9, billed itself as "Florida's Best Family Recipe!" The logo's annual berry character wore a chef's hat and was shown whipping a bowl of cream—likely to be used for shortcake, of course.

In the last several years, the queen's pageant had seen three history-making moments, and the 2014 pageant would bring yet another. Jessi Rae Varnum was crowned queen twenty-nine years after her mother, Kay Newsome, served as the festival's queen in 1985, making them the second mother/daughter duo to both hold the title. Queen Jessi Rae was selected with first maid Lindsey English and court members Macaley Barrow, Kallee Cook, and Caitlyn Kent.

A full cast of entertainers were on the playbill. Shoji Tabuchi, Styx, Ronnie Milsap, Colt Ford, Jimmy Sturr and Orchestra, the Tommy Dorsey Orchestra, Love and Theft, Little Big Town, Thompson Square, Rascal Flatts, Charley Pride, Josh Turner, Brenda Lee, Kellie Pickler, Crystal Gayle, Lee Brice, the Oak Ridge Boys, Third Day, John Anderson, Boyz II Men, Dustin Lynch, Jerrod Niemann, Easton Corbin, and the Band Perry gave festival goers plenty to choose from.

In these days, festival organizers were working arduously to bring fresh entertainment to the event, giving visitors something new to look forward to each year. This effort was successful in 2014 with these on-grounds additions: Wild About Monkeys, an educational show featuring trained baboons of all sizes; Redhead Express, a quartet of redheaded sisters, playing and singing with unbelievable harmony; the Walker Boys, brothers to the ladies of Redhead Express and talented, young bluegrass musicians; Savannah Jack, an up-and-coming country music trio; and special appearances by cast members of television's *Lizard Lick Towing* in addition to appearances by Troy Landry, R. J. Molinere, and Jay Paul Molinere from *Swamp People*.

A star of Wild About Monkeys performs with his trainer.

On Monday evening, Italy met Plant City when the festival's first strawberry spaghetti eating contest was set in motion. Contestants

The Redhead Express performs during their first festival visit.

Patrons peruse the Neighborhood Village's entries in its newly constructed building on the east side of the grounds.

were challenged with eating a half-pound bowl of spaghetti with a wooden spoon that was topped with . . . strawberries in place of meatballs.

The Neighborhood Village building construction crews worked arduously to finish by the opening of the 2014 festival and they were successful. The Neighborhood Village competition now had 4,000 square feet of space to display hundreds of entries. The horticulture show displayed its plants in its newly constructed space, and the Hall of Queens received a major redesign in its move from the main exhibit building. The space was fresh, air conditioned, and, best of all, allowed room for each of its three displays to grow.

2015

Two thousand fifteen brought a celebration to Plant City with the Florida Strawberry Festival's eightieth birthday. In deciding on a theme, organizers envisioned the eleven days as being one big party for the town and all of its special guests. And so, the theme for 2015 festival invited its patrons to "Come Join the Party!" from February 26 through March 8.

Samantha Sun was chosen to reign over the anniversary event with first maid Deanna Rodriguez and court members Payton Astin, Kellen Morris, and Emily Benoit. The festival even hired twenty-four of the best artists to entertain guests at the celebration: Bobby Vinton, Alabama, Mel Tillis, Jimmy Sturr and Orchestra, the Tommy Dorsey Orchestra, Scotty McCreery, MercyMe, Kevin Costner and Modern West, Brett Eldredge, Hunter Hayes, Sawyer Brown, the Happy Together Tour, Ronnie Milsap, Sara Evans, Ricky Skaggs, Craig Morgan, the Oak Ridge Boys, Newsboys, Loretta Lynn, Boyz II Men, Dan + Shay, John Legend, Parmalee, and Reba.

If visitors planned accordingly, they could be entertained all day with on-grounds acts. The Steve Trash Show, an educational show aimed at teaching the value of being environmentally conscious;

XPOGO, stunt men performing back flips and jumping over vehicles on pogo sticks; appearances from Troy, Chase, and Jacob Landry of *Swamp People*, and many more were spread throughout the grounds.

On the first Sunday evening, Major League Eating came to the festival with an event far outreaching shortcake, corn dog, or mashed potato eating. MLE contracted with the festival to host a qualifying event for the Nathan's Famous National Hot Dog Eating Championship that takes place every July 4 on Coney Island. What the festival thought would just be another eating contest actually drew competitors who traveled from Texas, California, and other places throughout the country. In the end, the winner of the male division downed twenty-three hot dogs and buns,

Troy, Chase, and Jacob Landry of *Swamp People* meet fans during their official meet and greet.

Eleven competitors from around the country compete in a Nathan's Famous Hot Dog Eating qualifying contest.

The owner of Stingray Chevrolet gets an extra hand in drawing the winner of a 2015 Chevrolet Corvette.

and the winner of the female division ate ten hot dogs and buns.

The eightieth festival came to a close with one lucky winner. Stingray Chevrolet gave away its sixth vehicle—a new Corvette.

In the spring of 2015, the board of directors selected Dan Walden to lead the governing body throughout the next two festivals.

2016

Since 1930, royalty has been a long-standing tradition at the Florida Strawberry Festival; a queen is crowned every year, Parkesdale Farms features a strawberry throne in its booth where visitors can sit for a photograph, and many refer to the festival's famous fruit as "King Strawberry." In 2016, the berry donned a crown and scepter of his very own for an event designed to be "Royal Fun for Everyone!"

But every king needs a queen. Haley Riley was chosen shortly before the festival to reign over the royal event with first maid Morgan Gaudens and court members Ashtyn Steele, Ashlyn Yarbrough, and Alex Aponte.

From the moment guests entered the grounds on opening day, March 3, they likely caught sight of a towering setup with rings and ropes that men

Mr. Berry greets guests at gate one on American Heroes Day.

and women would soon be hanging from. Circus Incredible, featuring members of the famous Wallenda family, made its debut at the festival in 2016. The group performed daring feats of balance and strength at jaw-dropping heights. In another show designed for younger audiences, the Belmont Magic Show awed children with tricks using rabbits, dogs, doves, and more. The country duo Branch and Dean made their festival debut, and the Redhead Express, Walker Boys, and Gothard Sisters all returned to loyal crowds.

When Mr. Belmont wasn't making things disappear on his magic show stage, he graciously hosted the new Fanta Strawberry Soda Throwdown. Competitors drank thirty-two ounces of the red drink as quickly as possible in hopes of winning a festival food certificate.

Festival organizers contracted with Charley Pride, Josh Turner, Mickey Gilley, Big & Rich, Lecrae, Cheap Trick, Shenandoah, Donny and Marie, Gene Watson, Lonestar, Trace Adkins, Ray Stevens, Martina McBride, the Oak Ridge Boys, Casting Crowns, Tanya Tucker (replaced by John Anderson), Cole Swindell, Merle Haggard (replaced by Kris Kristofferson), Charlie Wilson, echosmith, the Band Perry, and the Golden Boys starring Frankie Avalon, Fabian, and Bobby Rydell. All performed in the festival's headline entertainment shows.

After its closing on March 13, organizers dared not say the 2016 festival was perfect, but it was just about as close to perfect as a festival could get. Shortcake vendors sold 250,000 of the sweet treats; not one significant security threat took place; 560,487 guests joined in the festivities; and the weather was pristine, not a cloud in the sky and temperatures never rising much above ninety degrees. Somehow, Plant City maintained perfect weather for eleven days. Well . . . almost. The rains came in the last thirty minutes on the final night. But, at that point, no one could complain. General Manager Paul Davis said in a festival press release, "We had a phenomenal run, and we are truly thankful. We had terrific weather, good crowds, and everyone seemed to be enjoying time with their families." And that goal will continue to reign supreme for festivals to come.

Members of the famous Wallenda family, stars of Circus Incredible, perform death-defying acts.

AUSAGE ITALIAN
Livestock Arena
Livestock Arena
EcoShield

RENCH FRIES
RENCH FRIES
Fiske
SINCE 1938

FLORIDA
STRAWBERRY
FESTIVAL
Wish Farms
VISIT US TODAY TO SEE
HOW MUCH YOU COULD SAVE.

Wish Farms
Wish Farms
EXIT
VISIT US TODAY TO SEE
HOW MUCH YOU COULD SAVE.

APPENDIX I:
Florida Strawberry Festival Charter Members

H. M. Barker

G. H. Bates

Arthur R. Boring

G. A. Carey

Mrs. Henry Carlton

Marcus E. Cone

F. E. Cummins

John C. Dickerson

J. B. Edwards

H. A. Hampton

J. W. Henderson

James A. Henderson

H. H. Huff

Miss Motelle Madole

W. D. Marley

Miss Irene Merrin

H. S. Moody

P. M. Moody

T. E. Moody

G. O. Parker

G. R. Patten

F. M. Prewitt

Al Schneider

Wm. Schneider

Mrs. J. G. Smith

E. W. Wiggins

APPENDIX II:
Florida Strawberry Festival Queens and Courts

1930

Queen
Charlotte Rosenberg

Maids
Nettie Simmons
Alice Maxey
Kathryn Dudley
Irvin Wilder
Alice Sly

1931

Queen
Irvin Wilder

Maids
Bernice Adams
Gladys Balliette
Elizabeth Carey
Orel Ferguson
Elizabeth Hull
Genevieve McDermid
Elizabeth Morse
Eleanor Murrill

1932

Queen
Elizabeth Carey

Maids
Winifred Yates
Lorraine Parolini
Ethel Randall
Marian Herring
Dorothy Gentry
Elsie Cornett
Ruby Pratt
Verna Jones

1933

Queen
Christine Walden

Maids
Christine McDonald
Lillian Schulte
Mary May
Mildred Calding
Dorothy Adelson
Nona Mae Holloway
Anne Vannerson
Eloise Horton

1934

Queen
Dorothy Adelson

Maids
Goldie Rice
Doyle Colding
Clara Nell Register
Lila Florence Craig
Ermeldine Peeples
Edith Gentry
Elizabeth Buckalew
Helen Read

1935

Queen
Virginia Moody

Maids
Viva Lu Alexander
Mildred Sims
Christine McDonald
Elizabeth Hull
Mary Gay Head
Betty Rose Wright
Elizabeth Buckalew
Verna Roberts

1936

Queen

Pauline Schulte

Maids

Marzie McDonald
Illa Claire Chiles
Steevie Lee Wiggins
Catherine Henderson
Mary Frances Morse
Mary Ray Jones
Marie Register
Mildred Peacock

1937

Queen

Eloise Howell

Maids

Lauralou Harold
Marguerite Yates
Lucille McClellan
Catherine Fletcher
Ann Eberhardt
Virginia Dennison
Robbie Jarvis
Marceil Booth

1938

Queen

Norma Robinson

Maids

Betty Jo Cason
Nancy Simmons
Robbie Jarvis
Louise Youngblood
Susie Darrow
Thelma Waller
Donna Ve Walden
Doris Tomberlin

1939

Queen

Faye Mott

Maids

Lorene Forbes
Dollie Carter
Daphne Davis
Edna Ruth Peacock
Iris Curl
Mary Tillman
Betty Jo Cason
Christine Cassels

1940

Queen

Catherine Fletcher

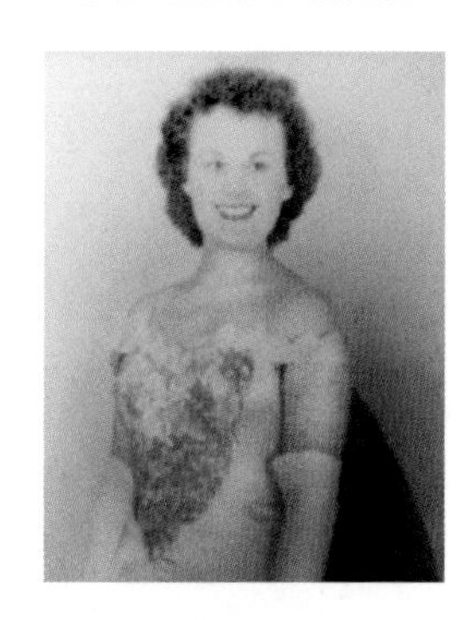

Maids

Mina Claire Gentry
Miriam Vannerson
Nettie Mae Berry
Helen Spear
Christine Cassels
Evelyn Downing
Ruth Gainey
Iris Curl

1941

Queen

Jane Langford

Maids

Helen Wells
Louise Gardner
Marjorie Rogers
Lillian Ergle
Virginia Jordan
Agatha Jean Sampson

1948

Queen
Barbara Alley

Maids
Dot Cone
Carolyn Campbell
Gloria Whilden
Eldora Holsberry

1949

Queen
Peggy Hodges

Maids
Peggy Sparkman
Carolyn Campbell
Jean Ann Ramsdell
Dot Cone
Sarah Jean Floyd
Addie Brown

1950

Queen
Etta Mae Helzer

Maids
Betty Chapman
Elyse Shuman
Inez Bender
Ann Adams
Joan Mabry
June Taylor
Joanne Pemberton
Betty Haney

1951

Queen
Edwana Snowden

Maids
Carol Kennedy
Gloria Hand
Irma Barta
Mary Lou Holloway
Letty Wilson
Margaret McLaughlin
Betty Sue Shuman
Mary Alice Etheridge

1952

Queen
Arvor Lois Harris

First Maid
Dottie Weldon

Maids
Ann Brown
Nell Jernigan
Elizabeth Sloan
Barbara Parish
Sandra Gentry

1953

Queen
Ruby Jean Barker

Maids
Eulene Brown
Patsy Phipps
Anna Clements
Dorothy Ann Shaw
Sandra Mays
Kathy Kimbel

1954

Queen
Ruth Shuman

Maids
Barbara Rowe
Shirley Patterson
Sandra Pritchard
Martha Lynn Ouellette
Ray Nell Simmons
Carolyn Newsome

1955

Queen
Betty Clements

Maids
Phyllis Hobkirk
Shirley Brown
Bobby Jean Harwell
Margaret Noe
Rosemary Hardy
Mary Jane Kennedy

1956

Queen
Virginia Young

First Maid
Mary Jane Jackson

Maids
June Fussell
Delores Harris
Carolyn Hardee
Maureen Crum
Marie Stanphill

1957

Queen
Linda Potter

Maids
Margaret Schenck
Judy Adams
Shirley Hutchinson
Dot Hardee
Wanda Harris
Elizabeth Bishop

1958

Queen
Sheila West

Maids
Carolyn Cunningham
Barbara Guice
Jackie Parker
Jeanette Sloan
Ida Lou Sapp
Yvonne Smith

1959

Queen
Lynda Eady

Maids
Joyce Whitehurst
Brenda Altman
Betty Saranko
June Ruis
Brenda Hagan
Gloria Veasey

1960

Queen
Betty Jean Cook

First Maids
Carolyn Duyck
Linda Duyck

Maids
Sylvia Jordan
Linda Wall
Joyce Eatman
Aloha Wills

1961

Queen
Diane O'Callaghan

Maids
Glenda Young
Cathie Mabry
Jackie Hardee
Patsy Whitehurst
Peggy Futch
Sylvia Whitehurst

1962

Queen
Marlene Coon

First Maid
Mona Le Cook

Maids
Mary Lynn Sapp
Diana Lott
Wanda Eady
Loretta Hay
Diana Lee Nagel

1963

Queen
Janice Barnes

First Maid
Vicki Murray

Maids
Jonadean Tramontana
Loretta Davis
Karen Dempsey
Sharon Young
Ola Jean Cason

1964

Queen
Georgie Blevins

Maids
Yvonne Valdes
Ava Faulkner
Sylvia Robbinson
Linda Herndon
Susan Kay McDonald
Monica Beach

1965

Queen
Sandra Link

First Maid
Linda Brown

Maids
LaVerne Ayscue
Sylvia Hagan
Cheryl Jones
Mary Frances Keene
Sheila Smith

1966

Queen
Kathy Mascali

First Maid
Linda Schweitzer

Maids
Pec Chambers
Sandra Leitner
Cheryl Barrington
Alice Mabry
Janie Goodell

1967

Queen
Maria Junquera

First Maid
Marcia Curl

Maids
Valera Stearns
Terry Gibbs
Karen Cobb
Lee Flag
Cherie Langland

1968

Queen
Sylvia Azorin

First Maid
Sharon Jeffries

Maids
Katha Head
Nancy Chaney
Gwen Johnson
Wendy Hester
Patti Moody

1969

Queen
Dee Zoet

Maids
Robyn Garrels
Yvonne Almon
Kathy Hockett
Sheila Howard
Judy Gay
Margie Tew

1970

Queen
Krysta Nifong

First Maid
Karen Campbell

Maids
Brenda Maggard
Maruchi Azorin
Tobe Robinson
Yvonne Chaney
Debra Williams

1971

Queen
Sherrie Chambers

First Maid
Donna Carpenter

Maids
Christy Cowart
Peggy Gardner
Barbara Cole
Cheryl Keene
Roseanna Massaro

1972

Queen
Linda Scanlon

First Maid
Carole White

Maids
Jenni Barwick
Crystal Beckum
Carol Tindall
Cathy Stowers
Ann Wynn

1973

Queen
Phyllis Head

Maids
Peggy Welch
Debbie Pollock
Melanie Frohlich
Cecille Dixon
Cindy Schipfer
Linda Adkinson
Sandra Lee Gollahon

1974

Queen
Denise Watts

First Maid
Mary Sloan

Maids
Lessie Werner
Lisa Sapp
Terry Archbell

1975

Queen
Sheryl Simmons

Maids
Martha Lastinger
Rebecca Pollock
Barbara Fulford
Cathy Johnson

1976

Queen
Martha Lastinger

First Maid
Karen Leitner

Maids
Barbara Fulford
Debbie Knight
Pamela Woods

1977

Queen
Karen Owens

Maids
Barbara Fulford
Amy Carpenter
Julie Willis
Molly Dull

1978

Queen

Mimi Phillips

Maids

Sharon Everidge
Cindy Copeland
Tammy Dukes
Kim Taylor

1979

Queen

Pam Smith

Maids

Lisa Freely
Patti Weyand
Karen Hall
Teresa Sluder

1980

Queen

Lisa Harris

First Maid

Tina Davis

Maids

Tami Napier
Holly Smith
Tamra Coton

1981

Queen

Pamela Farris

Maids

Annette Kilgore
Julie Adcock
Lisa Long
Nancy Wright

1982

Queen

Nancy Wright

Maids

Teresa Adams
Teresa Lott
Patricia Baskin
Scotti Ray

1983

Queen

Lisa Johnson

First Maid

Suzanne Patterson

Maids

Kim Deshong
Dee Newsome
Leslie Brown

1984

Queen
Sandra Howard

First Maid
Susie Hull

Maids
Donna Hodges
Roxanne Griffin
Melissa Pollard

1985

Queen
Kay Newsome

First Maid
Lillian Bowen

Maids
Michelle Griffin
Patty Harwell
Donna Hodges

1986

Queen
Rebecca Lewis

Maids
Stephanie Parker
Kerri Robinson
Pamela Sparkman
Teresa Tower

1987

Queen
Rebecca Brown

First Maid
Shayla Wetherington

Maids
Pamela Sparkman
Stacey Shearer
Jeanne Roundtree

1988

Queen
Robyn Simmons

First Maid
Dawn Dellapa

Maids
Christa Moore
Stacey Kendziorski
Daun Hollins

1989

Queen
Kristen Martinson

First Maid
Teresa Tower

Maids
Jennifer Kubler
Stacee Williams
Candice Moulton

1990

Queen
Joanna Cooper

First Maid
Beverly Forman

Maids
Michelle Nail
Lori McGinnes
Traci Bailey

1991

Queen
Stephanie Chesser

First Maid
Kim Hursey

Maids
Karessa Montgomery
Christy Holt
Patti Corley

1992

Queen
Lisa Diane Stanaland

First Maid
Stephanie Goff

Maids
Trisha Bailey
Monica Brock
Wendy Maggard

1993

Queen
Ashley Moody

First Maid
Julie Henderson

Court
Tonya Morrow
Rachel Smith
Jennifer Smith

1994

Queen
Amy Swilley

First Maid
Kellie Heth

Court
Cassandra Howard
Brittany Boothe
Ann Poonkasem

1995

Queen
Courtney Lea Clark

First Maid
Angela Williams

Court
Delana Hinson
Shelly Nickerson
Schyler Pickern

1996

Queen
Amy Norman

First Maid
Shea Wooten

Court
Emilie Dubois
Melonie Wilkerson
Shelley Causey

1997

Queen
Stephanie St. Martin

First Maid
Heather McDonald

Court
DeAnna Blount
Patricia Moody
Natasha Horn

1998

Queen
Jessica McDonald

First Maid
Amber Farmer

Court
Kathleen Guy
Kathie Heth
Emilie Dubois

1999

Queen
Kayla Drawdy

First Maid
Kristen Parke

Court
Jami Waters
Elizabeth Raburn
Charleene Closshey

2000

Queen
Alison Archbell

First Maid
Erin Welch

Court
Kristen Conte
Amanda Hall
Jennifer St. Martin

2001

Queen
Kellie Hinson

First Maid
Lindsay Roberts

Court
Ashley McDonald
Deanna Clemons
Evie Simmons

2002

Queen
Shannon Davis

First Maid
Amber Kosinsky

Court
Toinette Gerena
Holly Stein
Amanda Adams

2003

Queen
Erica Der

First Maid
Brandie Johnson

Court
Allison Bethea
Shana Norris
Kaley Mercer

2004

Queen
Kaitlin Sharer

First Maid
Ashley Pippin

Court
Crystal Wiggins
Lyndsey Terry
Leanna Blake

2005

Queen
Ashley Watkins

First Maid
Caycee Hampton

Court
Catie Walker
Brooke Ellis
Amy Stewart

2006

Queen
Hannah Hodge

First Maid
Ilene Chavez

Court
Megan Shelley
Kayla Gaschler
Julie Boback

2007

Queen
Summer Pippin

First Maid
Alexandra Watkins

Court
Tara Parker
Mackenzie Clark
Kyndle Hampton

2008

Queen
Kristen Smith

First Maid
Shaunie Surrency

Court
Britney Balliet
Amanda Sparkman
Jaclyn Raulerson

2009

Queen
Lauren Der

First Maid
Sara Beth Newsome

Court
Joclyn Emerson
Megan Cochran
Morgan Feaster

2010

Queen
Natalie Burgin

First Maid
TyLynn Eben

Court
Rachel Hallman
Ashlyn Robinson
Dayla Dementry

2011

Queen
Victoria Watkins

First Maid
Victoria Garren

Court
Summer Norris
Taelor Highland
Kori Lane

2012

Queen
Chelsea Bowden

First Maid
Calli Jo Parker

Court
Erica Kelley
Olivia Higgins
Chelsea Talavera

2013

Queen
Kelsey Fry

First Maid
Ericka Lott

Court
Maddy Keene
Madison Astin
Jamee Townsend

2014

Queen
Jessi Rae Varnum

First Maid
Lindsey English

Court
Macaley Barrow
Kallee Cook
Caitlyn Kent

2015

Queen
Samantha Sun

First Maid
Deanna Rodriguez

Court
Kellen Morris
Emily Benoit
Payton Astin

2016

Queen
Haley Riley

First Maid
Morgan Gaudens

Court
Ashlyn Yarbrough
Alex Aponte
Ashtyn Steele

2017

Queen
Drew Knotts

First Maid
Marlee Arn

Court
Caroline Brummer
Courtney Coton
Ariel Navarrete

APPENDIX III:
Florida Strawberry Festival Presidents

1930–1931
Albert Schneider

1932–1933
W. D. Marley

1934–1941
George Carey

1948–1949
Gerald H. Bates

1949–1973
E. N. Dickinson *

1973–1981
Jack Dempsey*

1981–1983
R. E. "Roy" Parke Jr.*

1983–1985
B. M. "Mac" Smith Jr.*

1985–1987
J. Albert Miles Jr.*

1987–1988
James L. Redman*

1988–1989
M. P. "Bud" Clark*

1989–1991
William D. Vernon*

1991–1993
Al Berry

1993–1995
Joe Newsome

1995–1997
Terry O. Ballard

1997–1999
J. D. Merrill

1999–2000
Ray Rollyson Jr.

2001–2003
Robert S. Trinkle

2003–2005
Kenneth M. Lucas

2005–2007
Johnny Dean Page

2007–2009
Gary L. Boothe

2009–2011
Mike Sparkman

2012–2013
Ronald Gainey

2014–2015
James "Jim" Jeffries

2016–2017
Dan Walden

* From 1949 through 1991, the Florida Strawberry Festival presidency also included the Hillsborough County Fair.

APPENDIX IV:
Florida Strawberry Festival Directors

Last	First	Span
Abbott	J. A.	Executive Committee 1933
Albritton	Mrs. J. M.	Executive Committee 1933
Alley	Fred	Director 1957–1968, Director 1971–1974, Associate 1970 and 1975–1977
Armor	J. O.	Associate 1971–1975
Ballard	Terry	Associate 1976–1984, Director 1984–2016
Barker	Hilton M.	Executive Committee 1931, Director 1932–1941
Bates	Gerald H.	Executive Committee 1933, Director 1934–1938, 1948–1949
Berry	Al	Director 1972–2014, Emeritus 2014–Present
Boles	Renita K.	Associate 2006–Present
Boothe	Gary	Associate 1993–1999, Director 2000–Present
Boring	Arthur R.	Director 1934–1941, 1948–1963
Bowden	Hilman	Associate 1990–1998, Director 1999–2015
Carey	George A.	Executive Committee 1931, Director 1932–1941
Carlton	Harry S.	Director 1966–1994
Carlton	Henry	Director 1964
Carlton	Mrs. Henry	Executive Committee 1931–1932
Carpenter	A. E.	Associate 1970
Carpenter	Carl	Associate 1983–1992
Carr	John D.	Director and Norman McLeod Post 26 Commander 1963
Cassels	J. Edwin	Executive Committee 1932–1933, Director 1939–1941
Chambers	Betty	Associate 1984–1995, Director 1995–2006
Clark	M. P.	Associate 1970, Director 1971–1989, Associate 1989–1991
Colson	Rodney	Associate 1983–1995, Director 1996–2001
Cone	Marcus E.	Director 1930, 1939
Cook	Denton L.	Director 1948–1965

Coton	Daniel	Associate 2003–2006, Director 2007–Present
Cummins	Frank E.	Executive Committee 1931, Director 1930, 1932–1939
Davenport	E. O.	Director 1975–1976, Associate 1977–1980, 1983–1989
Davis	James	Associate 2013–Present
Dempsey	Jack	Director 1951–1982
Dickinson	Elmer N.	Director 1948–1976
Dowdell	Mrs. R. S.	Executive Committee 1932
Draughon	Nettie Mae	Associate 1976–1997
Driggers	Doug	Associate 2007–Present
Edwards	Frank	Director and Norman McLeod Post 26 Commander 1965
Edwards	J. B.	Executive Committee 1931–1933
Forehand	Jane	Executive Committee 1932–1933
Gainey	Ron	Associate 1995–2001, Director 2002–Present
Galloway	Bruce	Associate 1978, Director 1979–2002
Gibbs	Louise	Director, 1969–1970, Associate 1971–1972
Goff	Francis D.	Director 1934–1941
Green	C. G.	Director 1934–1938, 1948–1955
Griffin	Osborne	Associate 1970
Hall	Floyd	Associate 1998–2006, Director 2007–Present
Hampton	Hugh A.	Executive Committee 1931–1933
Henderson	James A.	Executive Committee 1931–1933
Henderson	James W.	Director 1930–1941
Hinton	Dr. Charles	Associate 1984–1997
Hooper	McIntire	Director 1973–1997
Howard	Dr. Charles	Associate 1979–1983
Huff	Henry H.	Director 1930–1941
Hull	Edgar	Director 1948–1975, Associate 1976
Hull Jr.	Alex	Associate 1973, 1978–1982, Director 1974–1977
Isaacson	Sundle F.	Director 1948–1965

Jeffries	Jim	Associate 1995–2003, Director 2004–Present
Jones	David L.	Director 1967–1972, Associate 1973–1976
Keel	William	Associate 2000–2016
Knotts	Andy	Associate 2002–Present
Kuhn	Andrew	Director and Norman McLeod Post 26 Commander 1955
Kuhn	Joseph W.	Director 1957, 1966–1967, 1969, 1977–1978, Associate 1974–1976, 1979–2000
Langford	Louis T.	Director 1934–1938
Lastinger	Oscar	Associate 1970–1973
Liles	Woodie A.	Director 1965–1967, Associate 1970–1973
Lott	Rick	Associate 2007–Present
Lucas	Ken	Associate 1978–1993, Director 1994–Present
MacInnes	William. C.	Director 1964–1977
Madole	Motelle	Executive Committee 1931–1932
Marley	W. D.	Director 1930–1934
Martin	Sayde Gibbs	Associate 1992–2000
McClelland	William R.	Associate 1970–1994, Director 1995–2005
McDugald	M. A.	Executive Committee 1931–1932
McGinnes Jr.	W. D.	Associate 1998–2006, Director 2007–Present
Merrill	J. D.	Associate 1982–1987, Director 1988–2001
Merrin	Irene	Executive Committee 1931–1932
Miles	David	Associate 1998–2013
Miles Jr.	J. Albert	Director 1966–1987, 1990–1993, Associate 1988–1989
Moody	Judge James S.	Director 1948–1982
Moody	Henry S.	Director 1930–1933, 1939–1941, 1948–1966, 1971, Associate 1970, 1974–1977
Moore	Frank	Associate 1970–1976, 1978–1981, Director 1977, 1982–1988
Morgan Jr.	Lennox E. "Rat"	Director 1968–1970, Associate 1971–1975
Newsome	Joe	Associate 1980–1982, Director 1982–Present
Nifong	Dwight M.	Director 1966–1976, 1980, 1983–1995, Associate 1977–1979, 1981–1982

Nulter	Fred W.	Director 1948–1962
O'Day	R. W. Tom	Director and Norman McLeod Post 26 Commander 1960
Oszmanski	Evelyn	Associate 1975–1976, 1980, Director 1977–1979, 1981–1997
Page	Johnny Dean	Associate 1991–1998, Director 1999–2007
Page	Percy	Executive Committee 1931–1933, Director 1934–1938
Parke Jr.	Roy	Director 1967–1994
Parker	Kenneth	Associate 2013–Present
Patten	G. R.	Director 1935–1941, 1954–1965, 1968, Associate 1970–1972
Peace	Kenneth	Associate 1995–2001, Director 2002–Present
Poppell	Jon	Associate 2002–2015, Director 2016–Present
Purcell	Lani	Director 1976–1984, Associate 1985–1990
Redman	James L.	Director 1966–2006
Rogers	T. E.	Director 1934–1941
Rollyson	Rhett	Associate 2014–Present
Rollyson Jr.	Ray	Associate 1970–1989, Director 1990–2006
Rutherford	William E.	Director and Norman McLeod Post 26 Commander 1961
Salvato	Dr. Michael	Associate 2003–Present
Schmidt	W. O.	Associate 1973–1976
Schneider	Albert	Director 1930–1932
Scott	Jim	Associate 2015–Present
Sharpton	A. E.	Associate 1971–1972
Sharpton	Bill	Director 1966
Simmons	E. W.	Director 1948–1964
Simmons	William J.	Director 1950
Smith	Ercelle	Associate 1990–1996, Director 1997–1999
Smith	John Gordon	Executive Committee 1931–1933, Director 1939–1941
Smith	Mrs. J. G.	Executive Committee 1931–1932
Smith	W. Reece	Director 1934–1941, 1948–1965, Associate 1970–1973

Smith Jr.	B. M.	Director 1967–1995
Spann	Charles E.	Director 1948–1965
Sparkman	Chris	Associate 2006–Present
Sparkman	Michael	Associate 1994–2000, Director 2001–Present
St. Martin	John	Associate 1970–1994
Storms	Don A.	Director 1962–1963
Sullivan	David	Associate 1996–2005, Director 2006–Present
Swingley	C. D.	Director 1965–1970, Associate 1971–1974
Sytsma	Sandee Parke	Associate 1995–2005, Director 2006–Present
Taylor	C. Harold	Associate 1975–1977, Director 1977–1981
Thompson	Homer T.	Associate 1973–1975
Trinkle	Robert S.	Associate 1983–1993, Director 1994–2002
Truehart	Harold S.	Director 1948–1951
Tyner	John Porter	Director and Norman McLeod Post 26 Commander 1959
Varnum	Kay	Associate 2002–Present
Vernon	William D.	Associate 1974–1976, Director 1977–2001
Walden	Dan	Associate 1996–2002, Director 2003–Present
Walden	David C.	Director 1958
Walden	J. V.	Director 1977–1994
Waldron	Phillip	Associate 1998–2014, Director 2015–Present
Warner	Donald R.	Director and Norman McLeod Post 26 Commander 1962
Warnock	Pamela C.	Associate 2001–2014, Director 2015–Present
West	Harrell B.	Executive Committee 1933
Westfall	Charles R.	Director 1964–1967, Associate 1970–1972
Wetherington	Lane	Associate 2007–Present
Wilkes	Danny L	Associate 2001–2013
Winter	Peter	Director 1948–1964
Wright	C. P.	Executive Committee 1932–1933

BIBLIOGRAPHY

Publications

Bruton, Quintilla Geer and David E. Bailey, Jr. *Plant City: Its Origin and History*. Winston Salem: Hunter Publishing Company, 1984.

Gannon, Michael. *Florida: A Short History*. Gainesville: University Press of Florida, 2003.

Kerlin, Mark W. "Plant City, Florida, 1885-1940: A Study in Southern Urban Development." A thesis, University of Central Florida, College of Arts and Sciences, Department of History, Summer 2005.

Tidd, James Francis. "The Works Progress Administration in Hillsborough and Pinellas Counties, Florida, 1935-1943." A thesis, University of South Florida, 1989. Electronic version.

Vickers, Raymond B. *Panic in Paradise: Florida's Banking Crash of 1926*. Tuscaloosa: The University of Alabama Press, 1994.

Collections

Covington, Jane Langford and Harrison Wall Covington; Papers and Memorabilia.

Florida Strawberry Festival Association, Souvenir Programs, various from 1930 through 1969.

Hillsborough County Historic Resources Survey Report, *Bealsville*, October 1998.

International Independent Showmen's Museum, Riverview, Florida.

Plant City Photo Archives and History Center collections.

Snow, Panky; Papers and Memorabilia

Newspapers

Courier, also known as the *Plant City Courier*. Plant City, Florida. Tuesday, March 10, 1936; Friday, February 26, 1937; Friday, March 11, 1949; Thursday, March 4, 1965. Et al.

Enterprise, also known as the *Plant City Enterprise*. Plant City, Florida

Lakeland Ledger. Lakeland, Florida. Friday, March 8, 1963.

Tampa Daily Times. Tampa, Florida. Friday, February 27, 1953.

Tampa Tribune. Tampa, Florida. Tuesday, March 3, 1964.

Blog

Project VI: Florida Extension in the Era of Segregation, University of Florida/IFAS. February 13, 2014.

INDEX

.38 Special, 157
1984 Olympics, 130
4-H Club, 45, 66, 72, 95, 104, 126, 159
4-H Dog Show, 143, 144
A.C.E Junior Museum Pioneers, 95
ABC TV, 152
Adams, Amanda, 144, 182
Adams, Ann, 173
Adams, Bernice, 28, 171
Adams, Judy, 174
Adams, Teresa, 15, 178
Adcock, Julie, 113, 178
Adelson Field, 16, 35, 45, 46, 48, 49, 50, 53
Adelson, Dorothy, 32, 34, 35, 171
Adelson, Mr. and Mrs. Samuel, 35
Adelson, Samuel, 35
Adkins, Trace, 140, 144, 148, 152, 157, 160, 165
Adkinson, Linda, 100, 177
Ag in the Classroom, 129
agricultural center, 157, 159
AgVenture Farm Tour, 143
Air Supply, 158
Alabama, 136, 143, 162
Aldean, Jason, 152
Alderman, Minnie Lee, 25
Alexander Street, 44
Alexander, Viva Lu, 171
Allen, Duane, 112
Allen, Elsie, 31
Allen, Ira M., 24, 28, 31
Alley, Barbara, 50, 51, 52, 53, 158, 173
Alley, Mr. and Mrs. Fred, 51
Allstar Weekend, 157, 158
Almon, Yvonne, 89, 176
Alsobrook, Elizabeth, 25
Altman, Brenda, 174
American Heroes Day, 155
American Legion, 8, 17, 27, 17, 31, 34, 39, 45, 46, 48, 49, 51, 53, 54, 55, 56, 58, 59, 60, 61, 62, 63, 66, 67, 68, 69, 71, 72, 73, 74, 76, 78, 86, 116, 147
American Legion Auxiliary, 54
American Legion Home, 45, 49, 52, 55, 58, 61, 76, 78, 113
American Model Shows, 39
American Showman, The, 110
Amusement Business, 139
Anderson, Bill, 111, 117, 149
Anderson, John, 133, 135, 150, 161, 165
Anderson, Lynn, 102, 103, 109
Andrews, Horace, 80
Andrews, Katherine, 25
Andrews, Otis, 61, 62, 69
Apollo I, 85
Aponte, Alex, 164, 184
Appaloosa, 97
Arch Deal, 107
Archbell, Alison, 141, 181
Archbell, Terry, 101, 102, 177
Armed Occupation Act of 1842, 13
Arn, Marlee, 184
Arthur Boring Building, 99, 109, 116, 127
Askew, Reubin, 96, 103
Astin, Madison, 159, 183
Astin, Payton, 162, 184
Atkins, Rodney, 153, 158
Atlanta, 120
Atlantic City, 48
Augusta, Maine, 114
Avalon, Frankie, 122, 165
Ayscue, LaVerne, 175
Ayscue, Vernon, 89
Azorin, Maruchi, 95, 176
Azorin, Sylvia, 88, 90, 176
Badcock Furniture, 80
Bahamas, 122
Bailey, Razzy, 118
Bailey, Traci, 128, 180
Bailey, Trisha, 129, 180
Bakst, Lee, 156
Ballard, Terry, 11, 136, 138, 186
Balliet, Britney, 151, 183
Balliette, Gladys, 28, 171
Band Perry, 158, 161, 165
bank crash, 15
Baptist Hospital Fund, 51
Bare, Bobby, 154
Barker, Mr. and Mrs. E. C., 59
Barker, Mrs. Earl, 25
Barker, Ruby Jean, 59, 158, 173
Barnes, Janice, 78, 175
Barnes, Mr. and Mrs. P. H., 78
Barrington, Cheryl, 83, 176
Barrow, Macaley, 161, 184
Barta, Irma, 173
Bartow, 48
Barwick, Jenni, 98, 177
Barwick, Tom, 90
Baskin Robbins, 108
Baskin, Patricia, 115, 178
Bates, Gerald, 28, 31, 32, 35, 36, 39, 49, 51, 54, 185
Battery E, 116th Field Artillery, 27
Bay Ford Tractor Company, 95
Beach Boys Family and Friends, 140
Beach, Monica, 175
Beachcomber, 82
Bealsville, 13, 36
Bealsville Community Club, 95
Beckum, Crystal, 98, 177
Beef Breeder Banquet, 66
BeeLine Fashions, 90, 95
Beery, Donald, 63
Bellamy Brothers, 113, 147, 150
Belle City Amusements, 20
Belmont Magic Show, 165
Bender, Chester, 96, 97
Bender, Inez, 173
Benoit, Emily, 162, 184
Bentley, Dierks, 148
Bernardi Exposition Shows, 28
Berry, Al, 11, 73, 82, 83, 84, 86, 88, 115, 129, 130, 186
Berry, Nettie Mae, 172, 172
Beta Club, 27
Bethea, Allison, 145, 182
Bible Based Fellowship Cathedral Choir, 152
Big & Rich, 148, 150, 165
big freeze, 14
Bilirakis, Mike, 123

Bishop, Elizabeth, 174
Black, Clint, 147, 149, 157
Blackhawk, 139
Blackwood Brothers, 102
Blackwood, R. W., 109
Blain, Lester "Buddy," 58, 60, 61, 64
Blaine Act, 31
Blake, Leanna, 146, 182
Blevins, Georgie, 80, 82, 175
Blount, DeAnna, 137, 181
Blue Grass Shows, 56, 62, 66, 67, 71, 72, 76, 80, 83, 84, 85, 86, 87, 88, 89, 91, 95
Board of Trade, 14, 24
Boards, Bikes & Blades, 148
Boback, Julie, 148, 182
Bobby Hendrick's Drifters, 158
Boone, Pat, 100, 122
Booth, Marceil, 40, 172
Boothe, Brittany, 133, 180
Boothe, Gary, 150, 187
Boring, Arthur, 35, 37, 39, 41, 47, 48, 52, 80
Boring, Mrs. Arthur (Dorothy), 46
Bowden, Barbara Alley, 71, 127
Bowden, Chelsea, 158, 183
Bowden, Hilman, 58, 77, 90
Bowen, Lillian, 119, 179
Boyz II Men, 161, 162
Branch and Dean, 165
Branch Ranch, 98
Brandon, 58, 75
Brandon Area Art Club, 72, 77, 79, 83
Brandon Art Center, 95
Brant, Verl and Jean, 127
Brantner, Terri, 105
Brewer, Dr. Harold "Hal," 94
Brice, Lee, 154, 161
Britt, Floy, 36, 43, 45
Brock, Chad, 153
Brock, Monica, 129, 180
Brooks, Garth, 128
Brooks, Patsy, 126, 130, 131, 133, 137, 139, 152
Brorein, Carl, 28, 47
Brorein, William G., 28, 32, 47
Brown, Addie, 173
Brown, Ann, 58, 173
Brown, Eulene, 173
Brown, Jim Ed, 109, 153
Brown, Leslie, 116, 178
Brown, Linda, 175
Brown, Rebecca, 122, 179
Brown, Ruth Shuman, 122
Brown, Shirley, 61, 174
Brownie Troop 243, 63
Brummer, Caroline, 184
Bruton, James, 71
Bruton, Quintilla Geer, 80
Bryan Elementary, 51, 130
Bryan, Luke, 158
Bryant, Anita, 101, 102, 104
Buchman, Paul, 62
Buck Trout Puppet Show, 143
Buckalew, Elizabeth, 171
Buckingham, Rosemary, 90
Budweiser, 121
Budweiser's Clydesdales, 121, 135
Buffalo Bisons, 16
Burgin, Natalie, 154, 156, 183
Burney Elementary, 51
Burns, Haydon, 82
Bush, George, 119
Bush, George H. W., 9, 111, 129, 130
Business and Professional Women's Club (BPW), 26, 27, 32, 35, 89
Busk, Walter C. Jr., 32
Byrd, Tracy, 139, 147
C & S Paint Company, 95
Caccamisi, Barbara, 145
Cagle, Chris, 152
Cajun Country, 111
Calding, Mildred, 32, 171
Cameo Products, 95
Camp, Jeremy, 157
Campbell, Carolyn, 173
Campbell, Dr. Doak, 51
Campbell, Glen, 121, 137, 144, 152
Campbell, Karen, 95, 176
Capitol Theatre, 48
Capra, Frank, 34
Carey, Elizabeth, 25, 28, 29, 30, 171
Carey, George, 26, 28, 29, 30, 31, 32, 38, 47, 57, 58, 60, 61, 66, 70, 185
Carey, Mrs. George A., 30
Carlton, Doyle Elam, 30
Carlton, George, 32
Carlton, Governor, 30
Carlton, Harry, 99
Carlton, Mrs. Henry, 170
Carlton, Nell, 30
Carpenter, Amy, 105, 177
Carpenter, Donna, 97, 176
Carter, Dollie, 172
Carter, Rosalynn, 111, 112
Cash, Johnny, 117
Cash, June Carter, 117
Cason, Betty Jo, 172
Cason, Ola Jean, 175
Cassels, Christine, 172
Cassels, J. Edwin, 32
Cassels, Kenneth, 114, 116, 126, 130, 131, 133, 136, 137
Casting Crowns, 149, 160, 165
Causey, Shelley, 136, 181
Chamberlain High School, 80
Chambers, Betty, 118
Chambers, Dr. Frank, 98
Chambers, Pamela (Pec), 83, 176
Chambers, Sherrie, 97, 127, 176
Champion, The, 87
Chancey, Robert E. Lee, 31, 47, 49
Chaney, Nancy, 176
Chaney, Yvonne, 95, 176
Chang, Juju, 152
Chapman, Betty, 173
Chapman, Steven Curtis, 148
Charlie Daniels Band, 126, 128, 130, 135, 144, 147, 152, 158
Chavez, Ilene, 148, 182
Cheap Trick, 165
Checker, Chubby, 122, 157, 160
Cheerios, 146
Chesney, Kenny, 142
Chesnutt, Mark, 132, 136
Chesser, Stephanie, 128, 180
Cheyenne Stampede, 133
Chiles, Illa Claire, 172
Chinese Acrobats, 141

Circus Incredible, 165
Citgo, 95
Citizen of the Year (Outstanding Citizen), 80, 82, 84, 85, 86, 88, 89, 90, 94
City Hall Park, 55, 58, 59, 61
City of Plant City, 66, 95
Civic Club Luncheon, 90, 94
Civil War, 13
Civilian Conservation Corps (CCC), 16, 36, 38
Civitan Club, 36
Clambake Seven, 151, 152
Clark, Courtney Lea, 134, 180
Clark, M. P. "Bud," 123, 186
Clark, Mackenzie, 182
Clark, Ray, 137
Clark, Robert Julian III, 105
Clark, Roy, 120, 121, 122, 124, 136, 143, 154
Clark, Terri, 140, 142, 145, 148
Clearwater Mall, 108
Clements, Anna, 173
Clements, Betty, 61, 62, 174
Clemons, Deanna, 143, 181
Closshey, Charleene, 140, 181
Clower, Jerry, 102, 128
Cobb, Karen, 176
Coca-Cola Bottling Company, 43
Coca-Cola Jim, 38
Cochran, Megan, 152, 183
Colbert, Claudette, 34
Cold War, 54
Colding, Doyle, 171
Cole, Barbara, 97, 176
Collins, LeRoy, 63, 68
Colorado Strawberry Festival, 64
Colored Farmers' Agricultural Display, 43
Colored Home Demonstration Exhibit, 36, 43, 45
Columbia University, 30
Cone, Dot, 173
Cone, Fred P., 41, 45
Cone, Marcus, 24
Coney Island, 163
Conlee, John, 122, 157
Conner, Doyle, 72, 73, 76, 94, 96, 99
Connors, Frankie, 47
Conte, Kristen, 141, 181
Cook, Betty Jean, 71, 73, 175
Cook, Kallee, 161, 184
Cook, Mona Le, 175
Cooke, Kathryn, 86
Coon, Marlene, 76, 78, 175
Cooper, Joanna, 128, 180
Copeland, Cindy, 107, 178
Corbin, Easton, 157, 158, 161
Cork, 7, 24
Corley, Patti, 128, 180
Cornelius, Helen, 109, 115, 124, 153
Cornell Gunter's Coasters, 158
Cornett, Elsie, 171
Coronet Phosphate Company, 14
Corsair, 125
Coton, Courtney, 184
Coton, Tamra, 111, 178
Country's Reminisce Hitch, 136
County Association for Mentally Retarded Children, 95
County Council Homemaker's Extension Service, 95
Courier, 24, 25, 39, 46, 49, 53, 54, 56, 81, 86, 87, 88, 90, 101, 108, 109, 116, 130, 133, 135, 136
Covington, Harrison Wall, 49
Covington, Mrs. Harrison Wall, 51
Cowart, Christy, 97, 176
Cowboy Troy, 150
Cracker Corner, 109
Craddock, Billy "Crash," 109, 113, 118
Craig, Lila Florence, 171
Crocker, Marche, 84
Crum, Maureen, 63, 174
Cuesta, Henry, 138
Cumbee, Christie, 110
Cummins, Clyde Foley, 122, 124
Cummins, F. E., 24, 29
Cunningham, Carolyn, 174
Curl, Iris, 172
Curl, Marcia, 176
Currington, Billy, 154
Curtis, Ken, 102
Cypress Gardens, 64
Cyrus, Billy Ray, 132, 133, 135, 137, 140, 145, 152, 157
Dade County Youth Fair, 107
Dairy Costume Ball, 121
Dallas Cowboy, 132
Dan + Shay, 162
Daniels, Charlie, 126, 128, 130, 147, 152, 158
Daniels, Mrs. Charles, 25
Darrow, Susie, 172
Dating Game, The, 100
Dave & Sugar, 113
Davenport, E. O., 99, 113
Davis, Danny, 106, 111
Davis, Daphne, 172
Davis, Loretta, 175
Davis, Paul, 11, 152, 165
Davis, Shannon, 144, 182
Davis, Tina, 111, 178
Dean, Billy, 132, 137
Del McCoury Band, 145
Dellapa, Dawn, 124, 179
Dementry, Dayla, 154, 183
Dempsey, Jack, 78, 85, 87, 99, 185
Dempsey, Karen, 175
Dennison, Virginia, 25, 28, 40, 172
Denny the Clown, 119
Denny's Music & Comedy Show, 124
Der, Erica, 145, 152, 182
Der, Lauren, 152, 183
Desert Storm, 129
Deshong, Kim, 116, 178
DeVane, E. J., 29
DeVane, Frank, 15
Diamond Jubilee, 68
Diamond Rio, 132
Diamond Ron, 158
Dickens, Little Jimmy, 133
Dickerson, John C., 24, 29
Dickinson, Elmer N., 54, 54, 68, 72, 74, 76, 79, 83, 85, 87, 99, 185
Dickinson, Fred, 94
Diffie, Joe, 135
Dixon, Cecille, 100, 177

Do Do Sales & Hobby of Seffner, 95
DoBell, Frederick, 30
Dokkies, 42, 43, 45, 47
Doobie Brothers, 157
Dover, 7, 24, 45, 75, 76
Downing, Evelyn, 172
Dr. Hook & the Medicine Show, 115
Dr. Martin Luther King, Jr., Boulevard, 50
Dr. Pall Bearer, 101
Dramatic Order of the Knights of Khorassan (DOKK), 42, 43
Drane Field, 53
Draughon, Nettie, 107, 110
Drawdy, Kayla, 140, 181
Dubois, Emilie, 136, 138, 181
Dudley, Kathryn, 25, 171
Dukakis, Michael, 124
Dukes, Tammy, 107, 178
Dull, Molly, 105, 177
Dust Bowl, 34
Duyck, Carolyn, 70, 175
Duyck, Linda, 70, 175
Eady, Carolyn, 131, 175
Eady, Lynda, 71, 174
Eady, Wanda, 175
East Hillsborough County, 25, 29, 49, 50, 76
East Hillsborough County Chamber of Commerce, 14, 64, 66, 80
East Hillsborough Democratic Club, 95
East Hillsborough Historical Society, 115, 117, 125
Eastern Air Lines, 53
Eastern Star, 26, 101
Eatman, Joyce, 175
Eben, TyLynn, 154, 183
Eberhardt, Ann, 40, 172
Echosmith, 165
Edgar Hull Jewelry, 24
Educational Activities Center, 119
Edwards, J. B., 170
Eisenhower, Dwight D., 54, 70
Eldredge, Brett, 162
Electrical Exhibition Building, 84, 86
Electrical Exposition, 74
Elks Lodge, 56
Ellis, Brooke, 147, 182
Elston, Dick, 86, 88
Emerson, Joclyn, 152, 183
Energy Exhibit, 106
Energy Management Center, 106
English, Lindsey, 161, 184
Engvall, Bill, 154
EPCOT, 111
Ergle, Lillian, 172
Etheridge, Mary Alice, 173
Evans, Dale, 97, 98
Evans, Sara, 154, 162
Everidge, Cecil, 58, 64, 79
Everidge, Sharon, 107, 178
Exile, 124, 126
expo building, 147
Fabian, 122, 165
Fairchild, Barbara, 149
Faircracker, 124
Falling Arches, 107
Fanta Strawberry Soda Throwdown, 165
Fargo, Donna, 104, 113
Farm Bureau Kidney Fund, 106
Farmer, Amber, 138, 181
Farmers Market, 101
Farris, Governor, 79
Farris, Pam, 113, 114, 178
Fator, Terry, 146
Faulkner, Ava, 175
Feaster, Morgan, 152, 183
Felix Cavaliere's Rascals, 158
Fender, Freddie, 104
Ferguson, Orel, 28, 171
Ferris wheel, 16
Festus, 102
Fewell, Jack, 32
First Federal Savings & Loan, 95, 101
First National Bank in Plant City, 80, 84, 86
First Presbyterian Church, 107
Flag, Lee, 176
Fleming, Ken, 125, 126
Fleming, Tobe, 125, 126
Fletcher Ford, 46
Fletcher Motors, 95
Fletcher, Catherine, 25, 40, 45, 46, 172
Fletcher, Mrs. R. M., 25
Floren, Myron, 122. 124, 128, 138, 145
Florida A&M, 36
Florida American Legion, 37, 39
Florida Conservation Commission, 95
Florida Cooperative Extension Service, 36
Florida Department of Agriculture and Consumer Services, 100
Florida Exhibit float, 17
Florida Federation of Fairs, 124
Florida Federation of FFA, 95
Florida Forestry Service, 95
Florida Grower and Rancher Magazine, 74
Florida Solar Center, 106
Florida Southern College, 52, 53
Florida State Chamber of Commerce, 47
Florida State Energy Office, 106
Florida State Fair, 38, 40, 42, 99
Florida State Fair Association, 32, 47
Florida State University (FSU), 51, 53
Florida Strawberry Festival and Hillsborough County Fair, 95, 121
Florida Strawberry Festival Association, 7, 8, 11, 45, 49
Florida Strawberry Festival Community Booth, 38
Florida Strawberry Festival Fairgrounds, 45
Florida Strawberry Festival Fashion Show, 149
Florida Strawberry Growers Association (FSGA), 116, 117
Florida's Plant City, 114
Floyd, Sarah Jean, 173
Forbes, Lorene, 172
Ford, Colt, 161
Forehand, Jane, 31, 34, 37, 41

Foreigner, 160
Forman, Beverly, 128, 180
Fortner, Paula, 109
Four Aces, 137
Four Lads, 137
Fox Brothers, 136, 137
Foxworthy, Jeff, 153
Frances Wordehoff Sunday School Class, 43, 45
Frantz Oil Cleaner, 95
Freely, Lisa, 110, 178
Fricke, Janie, 118, 160
Friend, Bill, 11
Friend, Billy, 11
Frizzell, David, 117
Frohlich, Melanie, 100, 177
Frontiersman, 102
Fry, Kelsey, 159, 183
FSU Flying Circus, 106
Fulford, Barbara, 103, 104, 105, 177
Fulton, Greg, 123
Fussell, June, 63, 174
Futch, Peggy, 175
Future Farmers of America (FFA), 24, 25, 36, 45, 66, 72, 74, 104, 159
Future Farmers of Florida (FFF), 36
Future Homemakers, 66
Gable, Clark, 34
Gadabouts, 71
Gainer, Derrick, 132
Gainey, Ron, 157, 187
Gainey, Ruth, 172
Gaither Vocal Band, 157, 160
Galaxy Girl, 156
Galaxy Globe of Death, 156
Game Boy, 111
Garden Club, 85, 87, 89
Gardner, Louise, 172
Gardner, Peggy, 97, 176
Garrels, Robyn, 89, 176
Garren, Victoria, 156, 183
Gaschler, Kayla, 148, 182
Gasparilla Parade, 38, 40
Gatlin Brothers, 120, 124, 130, 147, 148, 153, 158
Gatlin, Larry, 111, 124, 147, 148, 153
Gatlyn Sisters Trio Music Making, 95
Gaudens, Morgan, 164, 184
Gay, Judy, 89, 176
Gayle, Crystal, 122, 123, 152, 158, 161
Gehrig, Lou, 34
Gentry, Dorothy, 176
Gentry, Edith, 171
Gentry, Mina Claire, 172
Gentry, Sandra, 58, 173
Gerena, Toinette, 144, 182
Gibbs, Louise, 94, 99, 101, 108, 109, 110
Gibbs, Terry, 176
Gilbert, Brantley, 160
Gill, Justin, 140
Gill, Vince, 130, 132, 140, 144, 145, 147, 158
Gilley, Mickey, 115, 120, 128, 152, 165
Gimby, Bobby, 102
Glaros, Ewana, 25
Glaros, John, 73
Glenwood Springs, Colorado, 64
Gloriana, 160
Goff, F. D., 35
Goff, Stephanie, 129, 180
Golden Boys of Bandstand, 122, 165
Goldsboro, Bobby, 106
Gollahon, Sandra Lee, 100, 177
Good Morning America, 152, 153
Goodell, Janie, 83, 176
Gooding, Rick, 128
Goodwill Ambassadors, 118
Gore, Al Jr., 123
Gosdin, Vern, 128
Goss, Brady, 156
Gothard Sisters, 156, 165
Graham, Bob, 111, 112, 123
Graham, Eddie, 102, 103
Graham, Mike, 102, 103
Grammar Schools' Parent Teacher Association, 42
Grampa Cratchet, 133, 136, 150
Grand Ole Opry Legends, 133
Grandpa Jones, 104
Granfield Avenue, 121
Grants Mall, 85
Great Depression, 7, 15, 16, 24, 31, 36, 40, 43
Great Freeze, 15
Great Society, 70
Greater Plant City Chamber of Commerce, 114, 131
Greco, Dick, 88
Green, C. G., 35
Greenwood, Lee, 120, 128, 137, 143, 149, 158
Gregory, Larry, 89
Griffin, Michelle, 119, 179
Griffin, Robert, 133
Griffin, Roxanne, 118, 179
Groupo Los Nenes, 150
Growers Building, 14
Guice, Barbara, 174
Guy Lombardo Band, 154, 157
Guy, Kathleen, 138, 181
Hagan, Brenda, 174
Hagan, Sylvia, 175
Haggard, Merle, 130, 165
Hall of Queens, 160, 162
Hall, Amanda, 141, 181
Hall, Karen, 110, 178
Hallman, Rachel, 154, 183
Hampton, Caycee, 147, 182
Hampton, H. A., 170
Hampton, Kyndle, 182
Hancock, Horace, 87, 90
Hand, Gloria, 173
Haney, Betty, 173
Happy Together Tour, 162
Hardee, Carolyn, 63, 174
Hardee, Dot, 174
Hardee, Jackie, 175
Hardy, Rosemary, 61, 174
Harold, Lauralou, 40, 172
Harrington, Bob, 106
Harris & Co., 141
Harris, Arvor Lois, 58, 173
Harris, Delores, 63, 174
Harris, Emmylou, 130
Harris, Fred and Lonnie, 58
Harris, Jack, 141
Harris, Lisa, 111, 178

Harris, Wanda, 174
Hart, Gary, 119
Harvey, Annie Ruth, 25
Harwell, Bobby Jean, 61, 174
Harwell, Patty, 119, 179
Haven, Ned, 28
Hay, Loretta, 175
Hayes, Hunter, 160, 162
Head, Katha, 176
Head, Mary Gay, 171
Head, Phyllis, 101, 177
Heath, Michelle, 110
Helzer, Etta Mae, 54, 55, 173
Henderson, Catherine, 172
Henderson, James A., 170
Henderson, James W., 24, 29, 31, 32, 35, 38, 41, 43
Henderson, Julie, 132, 180
Henderson, Sarah, 25
Henry, J. Myrle, 98
Henson, Haywood, 130
Herndon, Linda, 175
Herold, Bill, 87
Herring, Marian, 25, 171
Hester, Wendy, 176
Heth, Kathie, 138, 181
Heth, Kellie, 133, 180
Hichipucksassa, 13
Hickman, Kenneth, 114
Higgins, Olivia, 158, 183
Highland, Taelor, 156, 183
Hillsboro Bank, 83
Hillsboro State Bank, 48, 98
Hillsborough Community College, 85
Hillsborough County, 13, 14, 25, 77, 105
Hillsborough County 4-H Foundation, 106
Hillsborough County Budget Commission, 30
Hillsborough County Cowbelles, 110
Hillsborough County Fair, 130
Hillsborough County Fair and Florida Strawberry Festival, 76, 77, 78, 79, 81, 83, 85, 87
Hillsborough County Farm Bureau, 74, 105
Hillsborough County FFA Federation, 106
Hillsborough County FFA Foundation, 106
Hillsborough County Homemakers Council, 81
Hillsborough County Junior Agricultural Fair, 74
Hillsborough County Nutritional Service, 95
Hillsborough County Planning Commission, 107
Hillsborough Marketing Division, 97
Hilsman, Mabel, 25
Hinson, Delana, 134, 180
Hinson, Kellie, 143, 181
Hinton Farms, 114
Hinton, Chip, 116
Hixon, Curtis, 52
Hobkirk, Phyllis, 61, 174
Hockett, Kathy, 89, 176
Hodge, Hannah, 148, 182
Hodges, Donna, 105, 118, 119, 179
Hodges, Mr. and Mrs. William C., 52
Hodges, Peggy, 52, 53, 63, 173
Hola!, 154
Holcomb, Chris, 121
Holland, Spessard L., 48
Hollins, Daun, 124, 179
Holloway, Mary Lou, 173
Holloway, Nona Mae, 32, 171
Holsberry, Eldora, 51, 173
Holt, Christy, 128, 180
Home Demonstration Club, 24, 28
Home Demonstration Exhibit, 36, 45
Homemakers Extension Service, 91
Honey Festival Queen, 48
Hooker, Gerald, 98
Hooker's Department Store, 98, 102
Hootie and the Blowfish, 150
Hoover, Herbert Clark, 31
Hopewell, 7, 13, 24
Horn, Natasha, 137, 181
Horton, Eloise, 32, 171
Hotel Colorado, 64
Hotel Plant, 15, 29, 30, 31, 32, 35, 36, 39, 41, 43, 44, 45, 46, 47, 48, 52, 53, 54, 56, 57, 58, 60, 61, 65, 66, 70, 72, 75, 79, 83
Hough, Julianne, 153, 154
Houser, Randy, 160
Houston, David, 102
Howard, Cassandra, 133, 180
Howard, Sandra, 118, 179
Howard, Sheila, 89, 176
Howell, Eloise, 40, 41, 172
Howell, Sarah, 13
Hudson, Jody, 110
Huff, Henry H., 24, 27, 29, 35
Hull Armory Building, 147
Hull, Alex B. III, 82
Hull, Anne, 151
Hull, Edgar, 24, 80
Hull, Elizabeth, 25, 28, 171
Hull, Mrs. Edgar, 25
Hull, Sandy, 110
Hull, Susie, 118, 179
Hull, Virginia, 125
Human Cannonball, 128
Hungarian Liberty Club, 75
Hurley, Steve, 156
Hursey, Kim, 128, 180
Hutchinson, Janice, 63
Hutchinson, John, 56
Hutchinson, Shirley, 174
Impelliteri, Vincent, 53
Independent Strawberry Growers of Wimauma, 125
Indian River, 126
Internal Improvement Act, 13
It Happened One Night, 34
J. G. Smith FFA Chapter, 109
Jackie's Dance Studio, 75, 77, 84
Jackson Elementary, 51, 63
Jackson, Alan, 152, 160
Jackson, Mary Jane, 63, 64, 174
Jackson, Mildred Sims, 64
Jackson, Reverend Jesse, 119
James Ranch, 106
JaneDear Girls, 158

Jarvis, Robbie, 40, 172
Jaws, 94
Jeff and Sherri Easter with the Isaacs, 152
Jeffcoat, Gladys, 11
Jeffcoat, Harry, 11
Jeffries, James "Jim," 160, 187
Jeffries, Sharon, 176
Jennings, Waylon, 128
Jernigan, Nell, 58, 173
Jimmy Sturr and Orchestra, 157, 158, 160, 161, 162
Johnny J. Jones Exposition. 16, 24, 30
Johnson, Brandie, 145, 182
Johnson, Cathy, 103, 177
Johnson, Ellany, 131, 153
Johnson, Gwen, 176
Johnson, Lisa, 116, 178
Johnson, Peggy, 115
Johnson, Lyndon, 70
Johnson's Barbecue, 129
Johnson's Restaurant, 85, 90, 94
Jones, Cheryl, 175
Jones, Dale, 137
Jones, George, 121, 128, 143, 147, 149, 153, 157
Jones, Mary Ray, 172
Jones, Reverend Art, 152
Jones, Tom, 152
Jones, Verna, 171
Jordan, Sylvia, 175
Jordan, Virginia, 172
Judd, Naomi, 125, 126
Judd, Wynonna, 125, 126, 132, 133, 137, 147
Judds, 121, 126, 157
JUMP! The Ultimate Dog Show, 159
Jumping Frog Contest, 75
Junior Agricultural Fair, 64, 66, 76
Junior Chamber of Commerce (Jaycees), 57, 58, 60, 61, 62, 63, 64, 68, 69, 70, 71, 97
Junior Royalty Pageant, 105
Junior Woman's Club, 8, 67, 69, 70, 82, 85, 86, 87, 88, 89, 90
Junquera, Maria, 21, 86, 88, 176
K.C. and the Sunshine Band, 115
Kachunga and the Alligator, 133
Kahelin, Dick, 109
Kandu Magic Show, The, 158
Kanyuksaw, 15
Kari & Billy, 159
Keene, Cheryl, 97, 176
Keene, Maddy, 159, 183
Keene, Mary Frances, 175
Keith, Toby, 144
Kelley, Erica, 158, 183
Kendziorski, Stacey, 124, 179
Kennedy, Carol, 173
Kennedy, John F., 70
Kennedy, Mary Jane, 61, 174
Kent, Caitlyn, 161
Kershaw, Doug, 106, 115
Kershaw, Sammy, 139
Kevin Costner and Modern West, 162
Kiddie Korral, 135
Kids Greasy Pig Show
Kilgore, Annette, 113, 178
Killer Beaz, 153
Kimbel, Kathy, 173
Kirk, Claude, 85, 94
Kiwanis Club, 14, 27, 48, 56, 85, 87, 90, 91, 95
Kiwanis-Lions Luncheon, 53, 59, 61
Knight, Debbie, 104, 177
Knight, Gladys, 137
Knight, Mrs. J. Edgar Jr., 25
Knight, Peter O., 36
Knight, Ray, 111
Knights, 13, 95
Knights Griffin Road, 109
Knights of Pythias, 32, 42, 45, 47
Knotts, Drew, 184
Knox, Glenn, 72
Kocher, David Wayne, 105
Kool & the Gang, 153
Korean Conflict, 54
Kosinsky, Amber, 144, 182
Kristofferson, Kris, 165
Kubler, Jennifer, 125, 179
Kwik-Chek, 68
La Nueva Ilusion, 154
Lacoochee, 58
Lad & Lassie, 86
Lady Antebellum, 157
Lagasse, Emeril, 140
Lakeland, 27, 130
Lakeland Ledger, 74, 78
Lambert, Miranda, 152
Land Boom, 14, 15
Land Bust, 15
Landry, Troy, 161, 163
Lane, Cristy, 122
Lane, Kori, 156, 183
Langford, Jane, 17, 47, 48, 51, 172
Langford, L. T., 35, 49
Langford, Randy, 99
Langland, Cherie, 176
Larry Gatlin and the Gatlin Brothers, 120, 124, 130, 147, 148
Lastinger, Martha, 103, 104, 105, 177
Lawrence, Tracy, 144, 154
Lecrae, 165
Ledger, 110
Lee, Brenda, 128, 147, 153, 158, 161
Lee, Dennis, 124, 133, 136, 150
Lee, Ernie, 71, 107
Lee, Johnny, 115
Legend, John, 162
Legionnaires, 37, 41, 56, 59, 68, 84, 87, 90
Leitner, Karen, 104, 177
Leitner, Sandra, 83, 176
Lennon, Paul, 115, 117, 118, 128
Lewis Family, 109
Lewis, Rebecca, 121, 179
Liberace, 34
Link, Sandra, 82, 84, 175
Lions Club, 8, 15, 20, 24, 27, 32, 39, 48, 52, 56, 64, 65, 66, 68, 70, 71, 72, 74, 76, 79, 81, 83, 84, 85, 86, 87, 88, 89, 97, 98, 152
Lions Club Auxiliary, 52, 64, 65
Lions, Kiwanis, Rotary Club Luncheon, 79
Lithia, 7, 24, 95
Little Big Town, 20, 161
Little Richard, 150

Little Texas, 136
Lizard Lick Towing, 161
Lomison, Mrs. Harry, 25
Lonestar, 147, 148, 165
Long, Lisa, 113, 178
Los Sobrinos De La Lluvia, 154
Lott, Diana, 178
Lott, Ericka, 159, 183
Lott, Teresa, 115, 178
Lovato, Demi, 158
Love and Theft, 153, 161
Loveless, Patty, 136, 154
Lowry, Madie, 25
Lowry, Mark, 149, 153
Lucas, Kenneth, 146, 187
Luis, Jose, 154
Lumberjills, 158
Lynch, Dustin, 161
Lynn, Loretta, 117, 120, 128, 139, 140, 143, 162
Lynyrd Skynyrd, 154
M-B Equipment Inc., 95
Maas Brothers Department Store, 65
Mabrey, Lloyd, 158
Mabry, Alice, 83, 176
Mabry, Cathie, 175
Mabry, Joan, 173
Madole, Miss Motelle, 170
Maggard, Brenda, 95, 176
Maggard, Wendy, 129, 180
Magic of Lance Gifford and Company, 156
Main Street Ice Cream Parlor, 102
Major League Eating (MLE), 163
Mandrell, Barbara, 106, 126, 130, 133, 137
Mandrell, Louise, 117, 120, 139, 147, 158
Marion County, 30
Marley, W. D. , 24, 29, 30, 32, 185
Marschalk, Susan, 125
Marshall, Peter, 114
Martin, Sadye, 190
Martinson, Kristen, 125, 179
Mascali, Kathy, 83, 84, 86, 176
Masonic Hall, 26
Massaro, Paul, 125
Massaro, Roseanna, 97, 176
Mathews, Jack, 94
Mathias, Bob, 111, 112
Mattea, Kathy, 128, 135
Maxey, Alice, 25, 171
May, Mary, 32, 171
Mayo, Nathan, 30, 34, 35, 37, 39, 41, 42, 46, 53, 55, 56, 58, 59, 60, 61, 65, 67, 68, 71, 72, 73
Mays, Arden, 43, 44, 52, 78, 80
Mays, J. P., 28
Mays, Sandra, 173
McBride, Martina, 142, 145, 149, 160, 165
McCann, Susan, 128, 133, 137
McClellan, Lucille, 40, 172
McCoy, Neal, 136, 137, 139, 140, 143, 145, 147, 152, 160
McCreery, Scotty, 160, 162
McCrory's Five and Dime, 61
McCurdy, Jennette, All Star Weekend, 157
McDermid, Genevieve, 28, 171
McDonald, Ashley, 143, 181
McDonald, Christine, 32, 171
McDonald, Heather, 137, 181
McDonald, Jessica, 138, 181
McDonald, Marzie, 172
McDonald, Susan Kay, 175
McDowell, Ronnie, 153, 158
McElveen, Harry, 75
McEntire, Reba, 20, 121, 122, 123, 128, 158, 162
McGinnes Lumber Company, 95
McGinnes, Lori, 128, 180
McGinnes, Willard D., 80
McKay, Dave, 139
McKay, Donald Brenham "D. B.," 28
McKay, Minister V. Michael, 152
McLaughlin, Margaret, 173
McLeod, Norman, 49
Media Party, 113
Media Preview, 113
Medley, Bill, 154
Melody Booth Orchestra, 121
Memphis, Michigan, 122
Mercer, Kaley, 145, 182
MercyMe, 152, 162
Merrill, J. D., 186
Merrin, Miss Irene
Messina, Jo Dee, 142, 148
Meyer, Christy, 131
Miami Beach, 82, 84
Michaels, Bret, 160
Midnight Madness, 118
Mighty Bluegrass Midway, 135
Miles, J. Albert, 121, 185
Miles, Marion, 32
Miley, Gray, 25
Millsap, Marjorie, 120
Milsap, Ronnie, 107, 118, 120, 124, 132, 142, 145, 161, 162
Miss America Beauty Pageant Parade, 48
Miss America Pageant, 17, 48
Miss Florida, 48, 103
Miss Oklahoma, 101
Model Shows of America, 32
Molinere, Jay Paul, 161
Molinere, R. J., 161
Mondale, Walter, 119
Montgomery Gentry, 150
Montgomery, John Michael, 139, 147
Montgomery, Karessa, 128, 180
Moody, Ashley, 132, 180
Moody, Frank H., 80
Moody, Henry, 24, 29, 35, 52, 53, 55
Moody, P. M., 170
Moody, Patricia, 137, 181
Moody, Patti, 176
Moody, Stella, 25
Moody, T. E., 170
Moody, Virginia, 36, 37, 39, 171
Moore, Christa, 124, 179
Moore, Justin, 157, 160
Moore, Sam, 154
Morgan, Craig, 162
Morgan, Lennox E., 51, 59, 65
Morgan, Lorrie, 128, 153, 160
Morris, Gary, 126
Morris, Kellen, 162, 184
Morrison, Tilrow, 125
Morrow, Buddy, 99
Morrow, Tonya, 132, 180
Morse, Elizabeth, 28, 171

Morse, Mary Frances, 172
Mott, Faye, 44, 172
Mott, Harold, 139
Moulton, Candice, 125, 179
Mr. Berry, 131, 134
Mr. and Mrs. Tourist, 96, 98, 111
Mraz, Mike, 115
Mueller, Sherrie (Chambers), 127
Muglestones, 115, 117, 118
Murray, Anne, 130
Murray, Vicki, 175
Murrill, Eleanor, 28, 171
Nagel, Diana Lee, 175
Nail, Michelle, 128, 180
Napier, Tami, 111, 178
Nash Ramblers, 130
Nashville Brass, 111
Nathan's Famous Hot Dog Eating Contest, 163
National Guard Armory, 16, 45, 46, 54, 121
National Strawberry Queen Contest, 64
Nature Appreciation, 98
Navarrete, Ariel, 184
Neighborhood Village, 116, 147, 160, 162
Nelson, Rick, 102
Nelson, Willie, 148
Netherton, Tom, 138
Nettles, Anna Mae, 25
Nettles, Louise, 25
New Deal, 16, 31, 34, 36, 38
New Yorker, 151, 152
Newfield, Heidi, 154
Newman, Jimmy C., 111
Newsboys, 162
Newsome, Carolyn, 174
Newsome, Dee, 116, 178
Newsome, Joe, 132, 186
Newsome, Kay, 119, 121, 161, 179
Newsome, Sara Beth, 152, 183
Newton, Wayne, 130, 133, 134, 135, 140, 144, 147
Niagara Therapy Manufacturing Company, 95
Nichols, Emily, 159
Nichols, Joe, 149
Nickerson, Shelly, 134, 180
Niemann, Jerrod, 161
Nifong, Krysta, 95, 176
Nitty Gritty Dirt Band, 153
Noe, Margaret, 61, 174
Norman McLeod Post 26, American Legion, 17, 49, 51, 54, 60, 64, 70, 79, 83
Norman, Amy, 136, 181
Norris, Shana, 145, 182
Norris, Summer, 156, 183
Nulter, Fred, 54, 61, 64, 68, 72, 74
NWA Championship Wrestling, 102, 106, 107, 109, 111, 112, 113, 115, 117, 118, 120, 121, 122
Nygaard, Claudia, 133
O'Callaghan, Diane, 73, 175
O'Dwyer, William, 53
O'Neal, Lule, 105
Oak Ridge Boys, 112, 113, 121, 126, 128, 133, 139, 143, 144, 146, 147, 150, 154, 158, 161, 162, 165
Olson, Richard, 123
Olympic Games, 111, 136
Orange Festival Queen, 48
Order of the Eastern Star, 101
Oreo, 121
Osmond Brothers, 118
Osmond, Donny, 165
Osmond, Marie, 124, 165
Ouellette, Martha Lynn, 174
Outstanding Citizen, 80, 82, 84, 85, 86, 88, 90
Owen, Jake, 153, 158
Owens, Karen, 105, 111, 177
P&O Steamship Lines, 82
Pac-Man, 111
Page, Jane Day, 28, 32
Page, Johnny Dean, 148, 150, 187
Page, Percy, 32, 35
Paisley, Brad, 144, 145, 146
Palace, The, 147
Palm Beach Post, 145
Parent-Teachers' Associations, 32
Parish, Barbara, 58, 173
Parke, Kristen, 140, 181
Parke, R. E. "Roy" Jr., 113, 116, 117, 123, 185
Parke, Terri, 123
Parker, Calli Jo, 158, 183
Parker, G. O., 170
Parker, Jackie, 174
Parker, Stephanie, 121, 179
Parker, Tara, 182
Parkesdale Farms, 89, 95, 113, 114, 119, 130, 145, 164
Parmalee, 162
Parolini, Lorraine, 171
Pasco County, 45, 58
Pasco Shopper, 104
Patten, G. R., 54, 76, 79
Patterson, Gloria, 32
Patterson, Shirley, 174
Patterson, Suzanne, 116, 178
Paul Bunyan Lumberjack Show, 156
Peacock, Dr. Jack, 98
Peacock, Edna Ruth, 172
Peacock, Mildred, 172
Pearce, Webb, 71
Pearson, John, 133, 136, 146
Peeples, Ermeldine, 171
Pemberton, Joanne, 173
Peninsular Telephone Company, 32, 47
Pensacola, 13
Perkins, Carl, 128
Phillips Exotic Petting Zoo, 147
Phillips, Mimi, 107, 178
Phipps, Patsy, 173
Pickern, Schyler, 134, 180
Pickler, Kellie, 153, 161
Pied Pipers, 151
Pierce Manufacturing, 130
Pinecrest FFA, 95
Pinecrest Future Homemakers, 95
Pioneer Village, 113, 119, 129, 134
Pippin, Ashley, 146, 182
Pippin, Summer, 182
Pirates of the Colombian Caribbean, 156
Plant City Auto Supply Company, 95
Plant City Board of Trade, 24

Plant City Builders Model Homes, 85, 87
Plant City Chamber of Commerce, 64, 95
Plant City Church of God, 153
Plant City Commission, 121
Plant City Community Booth, 35
Plant City Courier, 100, 107, 131
Plant City Dolphins, 119
Plant City Enterprise, 24, 25, 27, 28
Plant City Farmers Market, 43
Plant City Federation of Garden Clubs, 70, 72, 74
Plant City FFA, 109
Plant City Fire Department, 130
Plant City Flower Show, 35
Plant City Garden Club, 77, 80, 83, 95
Plant City High School, 50, 51, 52, 56, 66, 90, 95, 99, 106, 137
Plant City High School Band, 51, 55, 57, 60
Plant City High School Orchestra, 65
Plant City Horseshoe Pitching Club, 83, 88
Plant City Industrial Park, 121
Plant City Junior Woman's Club, 68, 74, 80
Plant City library, 35
Plant City Observer, 110
Plant City Photo Archives and History Center, 7, 8
Plant City Plaza, 85, 87, 90
Plant City Police Department, 130, 131
Plant City Quarterback Club, 47
Plant City Recreation Department, 80, 86
Plant City Shopper, 94, 116, 118
Plant City State Farmers Market, 44, 46
Plant City Strawberry Festival and Hillsborough County Fair, 89
Plant City Strawberry Festival and Hillsborough County Junior Agricultural Fair, 74
Plant City Woman's Club, 72, 74, 77, 79, 83, 85, 87, 89, 101
Plant City Youth Association, 63
Plant, Henry B., 13, 35, 148
Platters, 158
Pollard, Melissa, 118, 179
Pollock, Debbie, 100, 177
Pollock, Rebecca, 103, 177
Poonkasem, Ann, 133, 180
Pope, Dick, 64
Potter, Linda, 65, 174
Powell, Sandi, 133
Pratt, Ruby, 25, 171
Prells Broadway Shows, 56
Presley, Elvis, 70
Prewitt, Fenton M., 24
Price, Ray, 150, 157
Pride, Charley, 115, 124, 128, 132, 143, 152, 157, 161, 165
Pritchard, Sandra, 174
Prophet, Ronnie, 133
Prophets, 98
Purcell, Lani, 120
Quarles, James, 61
Quarter horses, 76, 97
Rabbitt, Eddie, 107
Raburn, Elizabeth, 140, 181
Radio Station 97 Country, 136
Raggs Kids Club Band, 146
Rambler, 67
Rambo, Dottie, 111
Ramsdell, Jean Ann, 173
Randall, Ethel, 171
Rascal Flatts, 147, 161
Raulerson, Jaclyn, 151, 183
Rawlins, Ben, 44, 63, 70, 72, 79, 83, 85, 88, 90
Rawls, Lou, 139
Ray, Scotti, 115, 178
Rayburn, Donald, 57, 58
Raye, Collin, 136
Read, Helen, 171
Reardon, John and Therese, 132
Reaves, John, 111
Red Hat Society, 154
Redhead Express, 161, 165
Redman, James L., 81, 96, 122, 186
Redman, Ruby Jean Barker, 71
Reed, Jerry, 102, 107, 109
Register, Clara Nell, 171
Register, Marie, 172
Renfro Valley Show, 52
REO Speedwagon, 154
Reppies, 133
Retton, Mary Lou, 130
Revels Ford Tractor Company, 95
Reynolds, Debbie, 148, 154
Rhodes, Dusty, 106, 107
Rhodes, L. M., 34
Rice, Goldie, 171
Rickert, Bill, 85, 88
Riley, Haley, 164, 184
Riley, Jeannie C., 102
Rimes, LeAnn, 147, 149
Roadster, 148
Roauer, Mrs. William, 53, 63
Robbins, Elliot, 43
Robbins, Marty, 109, 115
Robbins, Ronny, 118
Robbinson, Sylvia, 175
Roberts, Lindsay, 143, 181
Roberts, Verna, 171
Robinson Family, 133, 136
Robinson, Ashlyn, 154, 183
Robinson, Kerri, 121, 179
Robinson, Norma, 43, 44, 172
Robinson, Tobe, 95, 176
Robinson's Paddling Porkers, 150
Robinson's Racing Pigs, 121, 133, 136
Rock-It the Robot, 156
Rodriguez, Deanna, 162, 184
Rogers, Kenny, 135, 142, 147, 150, 157
Rogers, Marjorie, 172
Rogers, Roy, 97
Rogers, T. E., 35
Rollyson, Ray "Rolly" Jr., 140, 186
Roosevelt, Franklin Delano, 31, 34, 38, 40
Rosenberg, Charlotte, 16, 25, 171
Rosenberg, Fannie Leibowitz, 25
Rosenberg, Sam, 25
Rounds, Wesley, 130, 131
Roundtree, Jeanne, 122, 179
Rowdy Rooster Puppet Show, 148

Rowe, Barbara, 174
Royal Crown Shows, 52, 56, 58, 59
Royal Palm Shows, 35, 36, 37, 41
Rucker, Darius, 154
Ruis, June, 174
Runa Pacha Indian World, 136, 146, 150
Rupp, Carl, 130, 131
Ruskin County Agent Horticulture Bees, 95
Russell, Johnny, 133
Ryals, John, 85
Rydell, Bobby, 122, 165
Sampson, Agatha Jean, 172
Sanchez, Eugenia, 25
Sanford, 13, 148
Sapp, Ida Lou, 174
Sapp, Lisa, 177
Sapp, Mamie, 25
Sapp, Mary Lynn, 175
Saranko, Betty, 174
Saturday Night Live, 94
Savannah Jack, 161
Sawyer Brown, 121, 128, 136, 162
Scanlon, Linda, 98, 177
Schenck, Margaret, 174
Schipfer, Cindy, 100, 177
Schneider Memorial Stadium, 61, 63, 65, 66, 67, 69, 71, 72, 73, 80, 114, 125
Schneider, Albert, 24, 26, 28, 29, 185
Schneider, Wm., 170
Schulte, Lillian, 32, 171
Schulte, Pauline, 39, 40, 172
Schultz, Fred, 94
Schweitzer, Linda, 83, 176
Scott, Helen and Billy, 122
Scruggs, Earl, 104
Sea Lion Splash, 158
Sea World, 111
Seaboard Air Line Railroad, 60
Seaboard Coastline Railroad, 97, 95
Seaboard System Railroad, 118
Seals, Dan, 124
Sears Roebuck and Company, 60
Sedaka, Neil, 139, 147
Seffner, 7, 24
Selmon, Lee Roy, 111
Seminole Lake Boulevard, 132
September 11, 144
Sharer, Kaitlin, 146, 182
Sharpton Chevrolet, 95
Shaw, Dorothy Ann, 173
Shaw, Hank, 139
Shearer, Stacey, 122, 179
Shelley, Megan, 148, 182
Shelton, Blake, 152, 160
Shenandoah, 130, 165
Sheppard, T. G., 113, 160
Sholtz, David, 35, 36, 39
Shoock, April, 110
Shuman, Betty Sue, 173
Shuman, Elyse, 173
Shuman, Mr. and Mrs. Foy A., 60
Shuman, Ruth, 60, 62, 174
Silms, Stephanie, 108
Silver Blossom Shows, 43, 45, 47
Simmons, Evie, 143, 181
Simmons, Helen, 25
Simmons, Nancy, 172
Simmons, Nettie, 25, 171
Simmons, Ray Nell, 174
Simmons, Robyn, 124, 179
Simmons, Sheryl, 103, 177
Simpson, Jessica, 153
Sims, Mildred, 171
Skaggs, Ricky, 120, 121, 124, 128, 142, 145, 162
Slaght, Lee, 56
Sloan, Elizabeth, 58, 173
Sloan, Jeanette, 174
Sloan, Mary, 101, 177
Sluder, Teresa, 110, 178
Sly, Alice, 25, 171
Smith Motors, 67
Smith, B. M. "Mac" Jr., 117, 145
Smith, Connie, 133, 137, 148, 153
Smith, Edna, 25
Smith, Ercelle, 110, 125
Smith, Holly, 111, 178
Smith, Jennifer, 132, 180
Smith, Kristen, 151, 183
Smith, Michael W., 147, 154
Smith, Mrs. J. G., 170
Smith, Pam, 110, 178
Smith, Rachel, 132, 180
Smith, Reece, 28, 35
Smith, Sheila, 175
Smith, Yvonne, 174
Smothers Brothers, 152, 154
Snider, Mike, 133, 136, 137
Snowden, Edwana, 51, 56, 173
Social Security Act, 38
Social Security Administration, 38
Sonny's Barbecue, 138, 139
South Carolina, 39
South Florida Baptist Hospital, 37
South Florida Fair, 35, 38
South Florida Fair Association, 32
Southern Living Magazine, 156
Southern Star Bluegrass Band, 108, 133, 136, 150
SouthWest Dairy Farmers Milking Show, 159
Sovine, Red, 71
Sparkman, Amanda, 151, 183
Sparkman, Mike, 153, 187
Sparkman, Pamela, 121, 122, 179
Sparkman, Peggy, 51, 173
Spear, Helen, 25, 172
Speer Family, 98
Spokesman, The, 15
Spooner, Bob, 79, 87
Springfield, Rick, 157
Springhead, 7, 13, 15, 24, 40, 45, 60
Springhead Girl Scout Troop 21, 63
Spurrlows, 126
Sputnik, 54, 70
SS *Florida*, 84
St. Augustine, 13
St. Clement Catholic Church, 102, 115, 120, 122, 125, 130, 145
St. Martin, Jennifer, 141, 181
St. Martin, Stephanie, 137, 181
St. Petersburg, 27
Stanaland, Lisa Diane, 129, 180
Stanphill, Marie, 63, 174
Star Wars, 94
State Farmers Market, 17, 31, 43, 44, 45, 119

State Farmers Market Board of Directors, 44
State Marketing Board, 30, 34, 45
State Revenue Commission, 54
State Road 574, 129
Statler Brothers, 104, 107, 124, 132, 142
Stearns, Valera, 176
Steele, Ashtyn, 164, 184
Stein, Holly, 144, 182
Steve Hall and the Shotgun Red Show, 160
Steve Trash Show, 162
Stevens, Ray, 126, 128, 153, 165
Stewart, Amy, 147, 182
Stingray Chevrolet, 156, 164
Stone Canyons, 102
Stone, Doug, 132, 133
Storms, Don A., 74
Stowers, Cathy, 98, 177
Strait, George, 122, 123, 128
Strawberry Ball, 8, 99
Strawberry Cloggers, 108
Strawberry Girls, The, 151
Strawberry Grower of the Year, 106
Strawberry Harmony, 114
Strawberry Investigation Laboratory, 15
Strawberry Marketing Commission, 95
Strawberry Patch Kiddie Korral, 135, 143
Strawberry Rock N' Dance, 114
Strawberry Song, The, 115
Strawberry Strollers Square Dancers, 83
Strickland, Margaret, 25
Stuart, Marty, 135, 140, 153
Sturr, Jimmy, 128, 137, 138, 145, 146, 148, 151, 154, 157, 158, 160, 161, 162
Styx, 161
Suez Resort Hotel, 84
Sugarland, 152
Summerfield, 30
Sun, Samantha, 162, 184
Sunshine Express, 113
Surrency, Shaunie, 151, 183
Sutton, Dorothy, 25
Swamp People, 161, 163
Sweat, Leola, 25
Sweeney Family Band Country Comedy Revue, 159
Sweethearts of the Rodeo, 124
Swift, Taylor, 20, 153
Swilley, Amy, 133, 180
Swindell, Cole, 165
Swingley, C. D., 81, 83, 85, 87
Switchfoot with One Republic, 154
Sylvia, 117
Sytsma, Sandee, 130, 144, 151
Tabuchi, Shoji, 161
Talavera, Chelsea, 158, 183
Tallahassee, 13
Tampa, 13, 15, 27, 31, 35, 38, 40, 42, 46, 47, 52, 77, 88, 148
Tampa Bay Hotel, 35
Tampa Board of Trade, 32
Tampa Daily Times, 59, 64
Tampa Electric, 36, 74, 75, 77, 79, 82, 83, 85, 87, 90, 106
Tampa Music Company, 95
Tampa Times, 114, 115
Tampa Tribune, 61, 79, 94, 112, 113, 114, 115, 116, 117, 120, 121, 123, 125, 130, 132, 133
Taylor Rental, 97
Taylor, Jay, 156
Taylor, John, 62
Taylor, June, 173
Taylor, Kim, 107, 178
Tempos, 90
Temptations, 136
Tennessee, 13
Terry, Lyndsey, 146, 182
Tew, Margie, 89, 176
Third Day, 153, 161
Thomas, Opal Clair, 25
Thomas, Wayne, 24
Thompson Square, 161
Thompson, James R., 105
Thompson, Josh, 157
Tickel, Randy, 63
Tillis, Mel, 20, 71, 106, 122, 136, 137, 142, 144, 147, 148, 153, 154, 160, 162
Tillis, Pam, 135, 136, 142, 147, 160
Tillman, Mary, 172
Tindall, Carol, 177
Tippin, Aaron, 137, 143, 147
TobyMac, 158
Tomberlin, Doris, 172
Tomlin FFA, 95
Tomlin Junior High School, 68, 121
Tomlin Middle School, 125
Tomlinson, Trent, 153
Tommy Dorsey Orchestra, 161, 162
Tompson, Stephanie, 110
Toothman, Lyndal, 109
Tower, Teresa, 121, 125, 179
Townsend Plan, 38, 39
Townsend, Dr. Francis E., 38
Townsend, Jamee, 159, 183
Townsendite, 38, 39
Tracy, Stephen, 131
Tramontana, Jonadean, 175
Trapnell, 7, 24
Travelling Bell Wagon, 133, 136, 146
Travis, Randy, 124, 135, 139, 143, 145, 150, 153
Tribunes, 98, 102
Trick Pony, 147
Trinkle, Robert, 84, 85, 144, 186
Tritt, Travis, 130, 152, 153
Tucker, Tanya, 118, 126, 128, 133, 145, 157, 165
Turkey Creek, 7, 24, 45
Turkey Creek Assembly of God, 110, 115, 125
Turkey Creek Booster Club, 95
Turkey Creek FFA, 70, 95
Turkey Creek High School, 78, 94, 98
Turner, David, 84
Turner, Josh, 150, 152, 154, 158, 161, 165
Twenty-first Amendment, 34
Twitty, Conway, 117, 118, 122, 126
U.S. Department of Agriculture, 53

Union Station, 14
Unity in the Community, 156
University of Florida Agricultural Research and Education Center, 131
University of Florida College of Engineering, 106
Urban Cowboy Band, 115
US Coast Guard, 96, 97
US Department of Energy, 106
US Navy, 68
US Navy's "Chuting" Stars, 118
US Olympic Training Center, 112
Valdes, Yvonne, 175
Valdosta, 97
Van Shelton, Ricky, 126, 133, 143
Vannerson, Anne, 32, 171
Vannerson, Miriam, 172
Varnum, Jessie Rae, 161, 184
Veasey, Gloria, 174
Vereen, Ben, 140
Vernon, William D., 126, 186
Vincent, Rhonda, 148
Vinton, Bobby, 128, 135, 145, 147, 152, 157, 160, 162
Violet Street, 132
Viscomi, Nick, 135
W. W. Mac Department Store, 54
Wade, Elvis, 107
Walden Chevrolet, 46
Walden, Christine, 32, 171
Walden, Dan, 164, 187
Walden, Don, 46, 47, 49
Walden, Donna Ve, 172
Walker Boys, 161, 165
Walker, Catie, 147, 182
Walker, Clay, 154
Walkowicz, Chester, 122
Wall, Linda, 175
Wallenda family, 165
Waller, Thelma, 172
Walt Disney World, 94
Warnell Lumber and Veneering Company, 14
Warren, Fuller, 52, 53
Warren, O. P., 28
Waters, Jami, 140, 181
Watkins, Alexandra, 182
Watkins, Ashley, 147, 156, 182
Watkins, Betty Barker, 17
Watkins, Victoria, 156, 183
Watkins, W. T., 28
Watson, Gene, 152, 160, 165
Watts, Denise, 101, 102, 177
Wauchula, 30
Waycross Express, 121
Weil, Kurt, 110
Welch, Erin, 141, 181
Welch, Peggy, 100, 177
Weldon, Dottie, 58, 173
Wells, Helen, 172
Werner, Lessie, 101, 177
West Grandstand Exhibits, 128
West, Dottie, 109
West, Sheila, 67, 174
West, Shelly, 117
Wetherington, Shayla, 122, 179
Weyand, Patti, 110, 178
WFLA Channel 8, 130
Whilden, Gloria, 173
Whitaker, Pat, 31
White, Bryan, 140
White, Carole, 98, 177
White, Hester, 25
White, Jack, 65, 67
Whitehurst, Joyce, 174
Whitehurst, Patsy, 175
Whitehurst, Sylvia, 175
Wiggins, Crystal, 146, 182
Wiggins, E. W., 170
Wiggins, Hazel, 25
Wiggins, Steevie Lee, 172
Wild About Monkeys, 161
Wilder Road, 107
Wilder, Calffrey LaFayette, 25
Wilder, Irvin, 25, 28, 171
Wilder, Joanna Singletary, 25
Wilkerson, Melonie, 136, 181
William Schneider Memorial Stadium, 61, 62
Williams, Angela, 134, 135, 180
Williams, Broward, 94
Williams, Debra, 95, 176
Williams, Don, 118, 130
Williams, Hank Jr., 158
Williams, Stacee, 125, 179
Williams, W. T., 28
Willie & Company, 133, 136
Willis, Julie, 105, 177
Wills, Aloha, 175
Wilson Elementary, 51
Wilson, Charlie, 165
Wilson, Gretchen, 149, 158
Wilson, Joe, 66, 67
Wilson, Letty, 173
Wilson, W. L., 44
Winter Haven, 27
Winter Strawberry Capital of the World, 14, 148
Winter Visitor Club, 85
Womack, Lee Ann, 142, 144
Woman's Club Flower Show
Woman's Club of Plant City, 8, 24, 28, 35, 43, 69
Woodall, Daisy, 25
Woods, Pamela, 104, 177
Woolridge, Reverend Joe, 125
Wooten, Shea, 136, 181
Works Progress Administration (WPA), 16, 36, 38, 43
World Books, 95
World War I, 15, 32, 48, 49
World War II, 7, 17, 48, 49, 50, 52, 53, 68, 155
World's Largest Strawberry Sundae, 102
Worley, Darryl, 148
WPLA Radio, 66, 68, 84, 88, 95, 115, 129
Wright Arcade, 24
Wright, Betty Rose, 171
Wright, Nancy, 113, 115, 178
WTMV Channel 32, 130
WTOG, 101
Wynette, Tammy, 107, 108, 120, 139
Wynn, Ann, 98, 177
XPOGO, 163
Yarbrough, Ashlyn, 164, 184
Yates, Marguerite, 40, 172
Yates, Viola, 25
Yates, Winifred, 171
Ybor City, 27
Yearwood, Trisha, 132
Yellow Fever, 15
Yoakam, Dwight, 160
Yoho, Don, 71

York, Laura, 130
Young, Chris, 157
Young, Dr. Calvin T., 44
Young, Glenda, 175
Young, Jason, 156
Young, Sharon, 175
Young, Virginia, 63, 174
Youngblood, Louise, 172
Youth Parade, 109
Zacchini, Hugo, 128
Zephyrhills FFA Band, 81
Zoet, Dee, 89, 90, 176

ABOUT THE AUTHORS

Lauren McNair has served as the public relations and media representative for the Florida Strawberry Festival since 2013. She holds a bachelor's of science degree in agricultural communication from the University of Florida and has a rich history with both agriculture and the Florida Strawberry Festival. Lauren's love for agriculture was instilled at an early age, growing up in her family's feed and farm supply store and cattle and citrus operations. Her passion for the industry only strengthened while a member of the National FFA Organization for seven years. Lauren began showing livestock at the festival at the age of three and exhibited animals every year until her high school graduation. She was selected Florida Strawberry Festival Queen in 2009 and represented the festival in over eighty appearances as its ambassador. Lauren currently resides in Plant City with her husband Andrew. She is an active member of First Baptist Church of Plant City and enjoys the outdoors, cooking, and raising chickens.

Gilbert Gott has served as the executive director of the Plant City Photo Archives and History Center since its inception in 2000. He has served as an adjunct faculty member of the History Department at Hillsborough Community College in Tampa, Florida, and has taught history and political science at the Dale Mabry and the Plant City Campus. He is a member of the National Council on Public History, American Association for State and Local History, Florida Historical Society, and the Society of Florida Archivists. Gil holds a BSFS from Georgetown University and received an MA in history from Indiana University of Pennsylvania. He pursued postgraduate studies at the University of Pittsburgh. He has written over one-hundred articles on Plant City's history. In addition to his interest in reading, writing, and teaching history, Gil is a US Coast Guard certified captain and is a partner in a charter service, sailing a thirty-two-foot Allied Seawind II ketch out of Apollo Beach.